NEW TECHNOLOGIES AND EMERGING SPACES OF CARE

New Technologies and Emerging Spaces of Care

Edited by
MICHAEL SCHILLMEIER
Ludwig-Maximilians-Universität München, Germany
and
MIQUEL DOMÈNECH
Universitat Autònoma de Barcelona, Spain

Routledge
Taylor & Francis Group

LONDON AND NEW YORK

First published 2010 by Ashgate Publishing

Published 2016 by Routledge
2 Park Square, Milton Park, Abingdon, Oxfordshire OX14 4RN
711 Third Avenue, New York, NY 10017, USA

First issued in paperback 2016

Routledge is an imprint of the Taylor & Francis Group, an informa business

British Library Cataloguing in Publication Data
New technologies and emerging spaces of care.
1. Medical innovations – Social aspects. 2. Home nursing – Technological innovations. 3. Telecommunication in medicine.
I. Schillmeier, Michael W. J. II. Domènech, Miquel.
362.1'028–dc22

Library of Congress Cataloging-in-Publication Data
New technologies and emerging spaces of care / by Michael Schillmeier and Miquel Domènech [editors].
p. cm.
Includes bibliographical references and index.
ISBN 978-0-7546-7864-9 (hardback : alk. paper)
1. Older people—Medical care—Technological innovations. 2. People with disabilities—Medical care—Technological innovations. I. Schillmeier, Michael W. J. II. Domènech, Miquel.
RA564.8.N49 2010
362.198'97—dc22
2010028234

ISBN 13: 978-1-138-25006-2 (pbk)
ISBN 13: 978-0-7546-7864-9 (hbk)

Contents

List of Figures and Table

Figures

Table

Notes on Contributors

Editors

Michael Schillmeier is a Schumpeter Fellow of the VolkswagenStiftung and teaches Sociology, Science and Technology Studies (STS), Disability Studies and Empirical Philosophy at the Department of Sociology at Ludwig-Maximilians-University/Germany. He received his PhD from Lancaster University/UK. He has written widely on the material dynamics of societal ordering and change, cosmo-political events, on bodies/senses and dis/ability, on care practices, on the societal relevance of objects and the heterogeneity of the social. He has currently published *Rethinking Disability: Bodies, Senses and Things* (Routledge) and with Joanna Latimer *Un/knowing Bodies* (Wiley/Blackwell).

Miquel Domènech is Senior Lecturer in the Department of Social Psychology at the Universitat Autònoma de Barcelona. Part of this book has been written as Visiting Fellow at CSISP (Centre for the Study of Invention and Social Process) in Goldsmiths College, University of London. His research interests cohere broadly in the field of science and technology studies (STS), with a special focus on the relationship between technical innovations and power relationships. At the Universitat Autònoma de Barcelona he leads GESCIT (Grup d'Estudis Socials de la Ciència i la Tecnologia), a group of research on STS.

Contributors

Elena Bendien is associated researcher at the University for Humanistics Utrecht, the Netherlands. She received her first PhD in linguistics at the State Pedagogical University of Rostov-on-Don, Russian Federation. Having moved to the Netherlands 11 years ago she is currently focusing on process philosophy as applied to social sciences. Her second PhD, which is in social and cultural psychology, is planned for June 2010. The thesis will be published under the title From the Art of Remembering to the Craft of Ageing: A study of the Reminiscence Museum at Humanitas, Rotterdam (2010). Elena's main fields of interests are: memory studies, ageing studies, the meaning of remembering at a later age, social aspects of (informal) care, identity issues in later life, new forms of non-therapeutic care and intercultural and gender diversity of ageing societies within a globalised world.

Søsser Brodersen is assistant professor at the Technical University of Denmark. She has a master's degree in Civil Engineering. She has been working with citizens and environmental issues and has strong engagement in the Science Shop of DTU.

Steven D. Brown is Professor of Social and Organizational Psychology at the University of Leicester. His research interests are around the mediation of social remembering across diverse settings. These include: commemoration of the 2005 London Bombings; personal, family and institutional recollections of childhood traumas and challenges; and self-archiving in virtual social networking. He is author of Psychology without Foundations: History, Philosophy and Psychosocial Theory (with Paul Stenner, 2009, Sage) and The Social Psychology of Experience: Studies in Remembering and Forgetting (with David Middleton, 2005, Sage).

Dr Alexandra Hillman is a research associate based in the Centre for the Economic and Social Aspects of Genomics (Cesagen) at Cardiff University. Her ethnographic research explores the assessment, care and treatment of older adults across a number of clinical settings. She is interested in the interrelations between the production of medical knowledge and practice and patients experiences of care. Her current work provides an analysis of ethical practice in caring for older people in acute trusts in the UK and she has recently secured a Wellcome Trust fellowship to investigate the ethical implications of the current drive towards early detection in dementia care.

Joanna Latimer is Professor of Sociology at Cardiff University School of Social Sciences and PI in the ESRC Centre for the Economic and Social Aspects of Genomics. She teaches social theory, and biomedical and cultural sociology. Joanna has published widely on genetics, care, identity, materiality, power and relationality. She is currently engaged in an ethnography of the social, cultural and ethical aspects of anti-ageing science and medicine. She is chair of the Cardiff Ageing, Science and Older People Network and the Culture Imagination and Practice Research Group.

Hanne Lindegaard is associate professor in User Oriented Design at the Technical University of Denmark. She has a master's degree in European Ethnology from the University of Copenhagen, and a Ph.D. from DTU. The combination of the two disciplines has given her Sorong cross-disciplinary experiences combining ethnographic field study methodology with engineering design approaches. Her research focuses are on the socio-material approach to design using the theoretical and methodological framework from STS and discourse analysis.

Daniel López is Lecturer of Social Psychology at Universitat Oberta de Catalunya. He is currently working on the implementation of new technologies in care settings like Home Telecare from an STS (specifically ANT) and a biopolitical perspective. His main areas of interest are: a) the emergence of new spatialities and temporalities of care; b) the emergence of new practices of caring and security due to the increasing importance of technologies of accountability; and c) the enactment of hybrid forms of autonomy and independence.

Peter A. Lutz is presently a PhD Candidate at the IT University of Copenhagen. His doctoral research explores movements in aging home care. This research is empirically grounded on ethnographic fieldwork in the Unites States and Sweden. Theoretical intersections between anthropology, science and technology studies (STS) and healthcare IT design inform his approach, as do several years of professional work experience in the IT industry.

Christine Milligan is Professor of Health and Geography and Director of the Centre for Ageing Research at Lancaster University. Her work focuses around the intersection between informal (family) care-giving, voluntarism and social welfare, particularly in relation to older people and the changing nature of home. She has written widely on the topic in books and refereed journals. Her latest book: There's No Place Like Home: Place and Care in an Ageing Society was published by Ashgate in 2009.

Maggie Mort is Reader in the Sociology of Science, Technology & Medicine, Centre for Science Studies, Lancaster University where she holds a joint post between the Dept of Sociology and the Division of Medicine. Her research interests include participatory ethnography and she has researched and published extensively on innovative health technologies; expertise in health practice and user and citizen participation in health services. She is currently coordinator of the EC FP7 project EFORTT (Ethical Frameworks for Telecare Technologies for Older People at Home).

Ingunn Moser is professor and dean of the Department of Nursing and Health Care, Diakonhjemmet University College, Oslo, Norway. She has published extensively on technology, disability, care and their politics for more than a decade. In her current research she has turned towards aging, elderly care, and dementia care in particular. Technologies, embodiment and subjectivity however still make central foci of her research. Her most recent publication is with Mol, Moser and Pols (2010) Care in practice: on tinkering in clinics, homes and farms (Bielefeld: Transcript Verlag). Moser is also affiliated with the EU-project "Ethical Frameworks for Telecare Technologies for Older People living at Home" (EFORTT).

Mark Paterson is Lecturer in Human Geography at the University of Exeter. Prior to that, between 2002 and 2006, he was Lecturer in Philosophy at the University of the West of England (UWE). His first book, Consumption and Everyday Life (Routledge, 2005), was followed by The Senses of Touch: Haptics, Affects and

Technologies (Berg, 2007), written as Visiting Fellow at Macquarrie University, Sydney, and with a grant from the Arts and Humanities Research Council (AHRC). He has published journal articles in literature, philosophy and social science journals, and received grants to look at robot skin (GWR) and the haptic modelling of prehistoric textiles (AHRC-EPSRC). Currently he is writing Seeing with the Hands: A Philosophical History of Blindness for Reaktion (Forthcoming in 2010) and, with Martin Dodge, co-editing the colleciton Placing Touch, Touching Space for Ashgate.

Paula Reavey is Senior Lecturer in Psychology at London South Bank University, UK. Her research interests are around embodiment, social remembering and feminist theory. Recent work includes edited volumes: Visual Psychologies (Routledge, forthcoming); Memory Matters: Understanding Contexts for Recollecting Child Sex Abuse (with Janice Haaken, Routledge, 2009); and New Feminist Stories of Child Sexual Abuse: Sexual Scripts and Dangerous Dialogues (with Sam Warner, 2003); as well as a number of articles on child sexual abuse, sexuality and embodiment, using discourse analysis, visual methods and memory work. She is Associate Editor of the Psychology of Women Section Review.

Research Centre for Shared Incompetence / Xperiment! (**Bernd Kraeftner, Judith Kroell, Gerhard Ramsebner, Leo Peschta, Isabel Warner**) is a trans-disciplinary research group based in Vienna. The members of the working party contribute (in)competencies and skills in fine and digital arts, filmmaking, sociology, medicine, science and technology studies etc. The members share an interest in experimenting with scientific ideas at the messy intersection of the sciences, health care, politics, publics and the arts and have accomplished several research/art projects since 1998 (http://www.sharedinc.net/). Recent projects of the group are "Pillow research. Multiple diagnoses and hidden talents" and "In the event of ... Anticipatory and participatory politics of emergency stocks" (in collaboration with M. Guggenheim) focusing on provision and care in the event of anticipated emergencies funded by the Vienna Science and Technology Fund.

Celia Roberts is a Senior Lecturer in the Department of Sociology, Lancaster University and is a member of the Centre for Gender and Women's Studies and the Centre for Science Studies. She works on a range of health technologies, including telecare, reproductive technologies and pharmaceuticals. She is also currently involved in an EU-funded project on patient organisations and knowledge society.

Hilde Thygesen, trained as an Occupational Therapist and Sociologist is a researcher at the Diakonhjemmet University College of Oslo, Norway. For several years she has been working on issues of technology and care; as a heath care professional, in policy and as a researcher. Her main research interests are empirical ethics, and in particular questions and issues related to care practices and the use of technology.

Chris Tweed is Director of the BRE Centre for Sustainable Design of the Built Environment (SuDoBE) in the Welsh School of Architecture at Cardiff University. The main focus of his research is to discover how the built environment and its technologies can be designed to improve the quality of people's lives. His work employs socio-technical concepts to study the use of technology in people's homes, the creation and maintenance of thermal conditions in the home, sustainable design methods and design interventions to create sustainable communities. In addition to theoretical insights, the results of this research are intended to inform the design of more sustainable neighbourhoods, buildings and comfort systems.

Paul White is employed as a Research Associate within the Business School of Imperial College London, following a previous appointment in the Medical School of Cardiff University. He survived undergraduate, graduate training and clinical work in nursing and health care (in what was the University Of Wales College Of Medicine) and undertook further graduate and doctoral training in sociology (Cardiff University, School of Social Sciences). His research interests surround the tricky issue of 'stuff people do' and things that mediate accomplishment (he also considers himself an ethnographer) and appears to overuse brackets (like this one). For one reason or another Paul's research has generally been related to health care, although he considers this just another space where stuff happens. He has performed research in hospital intensive care units, primary care settings and his current research work is focussed around health care organisations, innovation and cultural change programs within and across health care institutions. He can often be found reading (or at least looking at) books.

Acknowledgments

The editors would like to thank all the contributors of this book for their commitment and participation in rethinking care practices. We would also like to thank the Spanish Ministerio de Educación y Ciencia (CSO2008-06308-C02-01), the European Commission (Project Number 217787/FP7-SCIENCE-IN-SOCIETY-2007-1) and the Deutscher Akademischer Austausch Dienst (DAAD D/06/12861). Thanks to their financial support we were able to meet face-to-face and work closely on the ideas captured in this book. Miquel Domènech would like to thank Mike Michael and the Centre for the Study of Invention and Social Process (CSISP) for their support while he was a Visiting Fellow at Goldsmith College, London.

Finally, the German artist Yvonne Lee Schultz has done it again and created wonderful cover-artwork. Thanks Yvonne!

New Technologies and Emerging Spaces of Care – An Introduction

Michael Schillmeier and Miquel Domènech

It has become commonplace to say that we live in a technological world. Technology is not only vital for producing societal progress but also for dealing with its negative side effects. Technology shapes our everyday life. If human beings lack the ability to perform in a socially 'normal' way, technology helps to fill the gap of normality. Hence, technology is a key actor in societal processes of normalization (cf. Schillmeier 2010).

A striking example is the practice of care (cf. Latimer 2000; Latimer and Schillmeier 2009; Schillmeier and Domènech 2009; Schillmeier and Heinlein 2009). Care for dependent people has become one of the major problems of ageing western societies. Policy makers are worried by future scenarios, where a lack of provision is envisaged for long-term care for an increasing population of elderly people who are no longer capable of taking care of themselves. There are two main factors that explain this problem. First, inversion of the population pyramid, due to a decrease in births combined with an increase in lifespan. Second, a progressive transformation, over the last fifty years, of the traditional model of the family, which is expressed in the incorporation of women in the labour market and the problematization of their role as prime informal caregivers at home (Outshoorn 2008).

Consequently, care, even in its more private forms, is becoming a highly regulated realm as well as the focus of different kinds of institutionalized interventions in order to provide a societal answer to the 'lack of care'. In effect, current health and health care systems are experiencing and dealing with complex transformations and varied reforms (Magnusson, Hanson, Brito, Berthold, Chambers and Daly 2002, Pickard 2009). Obviously, new technologies do play a key role in elaborating and implementing such reforms. For example, the British Government policy, as expressed in the Preventative Technology Grant, seeks to guarantee an almost universal telecare service for older people at home (Roberts and Mort 2009).

Technological and scientific innovations associated with solving the 'problem of care' are expected to produce societal transformations. They question and alter common social relationships and evoke, as well as stabilize, a new ordering of everyday life. As we well know, processes of social change induce a considerable reconfiguration and intermingling of our private and public spatial arrangements. New care and care systems technologies, such as telecare (Percival and Hanson 2006; Williams, Doughty and Bradley 2000), smart technologies and smart homes (Harper

2006; Soar and Seo 2007; Van Berlo 2002) shape and reshape the practices and spatialities of care (Milligan 2001; Poland, Lehoux, Holmes and Andrews 2005).

Needless to say, these changes are not exempt from controversies. Technology is considered helpful, but it is also criticized for undermining 'authentic' care delivery (Savenstedt, Sandman and Zingmarkk 2006). These concerns are certainly very common within western societies. The ideology of technology and its powerful effects are thought to be colonizing human interaction and reifying human relations (cf. Habermas 1985). This critique of technological rationalization relies on the idea that human society is about rational human beings and their mutual relationships. Technology, on the other, is seen as an 'angewandte Natur' (Marx) that may dominate human relations by rationalizing and objectifying them. However, from such logic, it remains difficult to understand the complexities of human society if we consider it a merely human affair. Unfortunately, the technophobia of social sciences' critique of modernity has been very much part of rendering human society fundamentally different from non-human, nature etc. (cf. Latour 2005). Michel Serres (1994, 1995) has stressed that social sciences have been careless when trying to explain society in terms of a *social contract*, which only binds human beings, and merely considers language, writing and logics. Things, devices and utensils have always played an important role in the constitution of human society.

It is in Science and Technology Studies (STS) where Michel Serres' proposals have gained evidence. Bruno Latour's idea that technology and devices are 'the missing masses' of societal explanations has became a central focus of social scientific research (Latour 1992). According to Latour, it is very difficult to build stable societies with humans alone. Rather, human societies achieve their solidity precisely by relying on the non-human, on technology. In that sense, technology has always been a way of solving the problem of building durable societies on a large scale (Strum and Latour 1999).

When sharing such an understanding of human society, it makes little sense to argue that technology interferes with care or 'colonizes' human care practices. Having said this, we are not arguing that technology is a neutral set of tools that is applied by care practices. Technology is not essentially neutral or just good or bad. To argue generally for or against technology in care practices is deeply floored. It is equally problematic to pretend that technology is neutral with effects that only depend on the intentions of the human actors who use it.

This book offers a different approach to technology, inasmuch as technology is seen as crucial in mediating human society and with it the questions concerning care practices. It is the main objective of this edited volume to provide empirically and conceptually rich insights into the different ways technology (re-)maps human society and its practical spaces of care. Through the poly-theoretical lens of Science and Technology Studies (STS), the contributions of this book provide a critical understanding of contemporary practices of care that cut through the growing conceptual and methodological barriers between social and medical models of care studies. To do so, it was apparently vital to draw upon context-specific, sociocultural relations (and changes) between human beings and non-human technology, artefacts and objects within health care practices. These practices give special attention to the materiality of every day social practices, i.e. how human beings and technology fabricate societal relations that disable and/or enable, question and/or alter, as well as stabilize new health care spaces. Moreover, it brings to the fore social and cultural dynamics. These arise from the introduction of (assistive) technology in the home and within public settings, including new forms of social control, resistance to, or acceptance of (assistive) technology: Sophisticated medical technology that used to be the exclusive domain of hospitals is becoming increasingly available in the home (Fex, Ek and Söderhamn 2009; Mann 2001). At the same time, more and more new technologies are being developed in order to obtain the same quality of care in care homes or hospitals as performed in private homes (Carrillo, Dishman and Plowman 2009).

This edited volume offers detailed studies of transnational and transdisciplinary research projects that explore and analyze the social and cultural impact of new technologies and how technological innovation configures and reconfigures institutionalized spaces of care. It highlights the societal relevance of new selected care technologies, and how technological and scientific innovations are closely associated with a number of highly contesting and contested societal affairs. With this in mind, *New technologies and emerging spaces of care* deals with, at least, four basic concerns that merit further consideration.

First, the effects of social control, accountability, surveillance and discipline associated with the introduction of technological advances, either at home or within institutionalized settings. In this sense, the influence of Michel Foucault's analysis on power is crucial for reflections on care (Vaz and Bruno 2003) and acquired a very specific discourse about the effects of technologies of care in care receivers (Essén 2008; Laviolette and Hanson 2007) and care workers (Hjalmarson 2009). It is important to note that the transition of care from institutions to the community does not mean the interruption of surveillance practices. As Bloomfield and McLean (2003) have explained very well, people outside the institutions are also rendered visible and subjected to practices of surveillance and control.

Second, the re-arrangements of relations between private and public forms of care. Probably one of the most striking characteristics of emerging spaces of care is the alteration of traditional boundaries between institutional and non-institutional spheres. This is particularly noticeable in understanding 'home', where it gives rise to different processes of boundary work oriented towards maintaining control over the home environment, integrity of self and assurance of personal safety (Dyck, Kontos, Angus and McKeever 2005). In effect, these re-arrangements are part of new individual as well as societal (economic, political, social etc.) practices and concerns (Nicolini 2006; Shine 2004).

Third, the generation of new relations and subjectivities of care receivers and care givers. Indeed, according to various research outcomes, it is possible to discern several transformations affecting the patterns of interaction in the clinical encounter among patients and professionals (May, Finch, Mair and Mort

2005; Mort, May and Williams 2003). Such interactional processes are difficult to understand without addressing the affective relations between people and technologies on the one hand, and the social relations that technology generates on the other (Pols and Moser 2009).

Fourth, the possible ethical implications of changes in care delivering due to the introduction of new technologies. What constitutes good care is certainly affected by the use of technology (cf. Mol 2008). The issue of technology adds a new dimension to the ongoing discussion about the meaning of autonomy in the realm of care ethics (Sybylla 2001; Verkerk 2001). New dependencies that accompany technological innovations generate a discourse on autonomy not as property of human beings but as the effect of human/non-human interdependencies that enable/disable certain possibilities of action, agency and discretion (López and Domènech 2009; Schillmeier 2008). Ethical concerns are especially relevant when the care receivers are limited in their capacity to express themselves as in the case of dementia. Consent, then, is a serious problem (Marshall 1999). This is precisely what has led some scholars to advocate specific ethical guidelines in order to help caregivers to be ethically responsible in their work (Magnusson and Hanson 2003).

Following from that, new technologies not only affect emerging spaces of care that question and alter traditional, institutionalized boundaries of care spaces and health care systems, they also create and demand new skills, new ways of experiencing the practices of everyday life (Loe 2010). Technology creates new risks and opportunities, not only for care receivers, but also for caregivers (Chambers and Connor 2002; Magnusson, Hanson and Nolan 2005). Moreover, we argue that the empirical complexities of technological innovation and new emerging spaces of care stipulate innovative, practice-oriented accounts that develop and refine methodological and theoretical tools for the analysis of the interaction of (assistive) technology with elderly and disabled people in their everyday life spaces. Thus, this book aims to provide insights into ways of supporting the (re-)shaping of health care policy and practice around the introduction of (assistive) technology for the elderly and disabled people. Furthermore, it also puts into practice novel concepts to adequately address the relationship between disabled, vulnerable, volatile people and technology and emergent health care system spaces and practices.

Therefore, the following chapters of this book share a practice-oriented attitude that tries to capture the often fragile, contradictory, contingent and multiple dynamics of embodied human life that is very much mediated by technoscientific objects and practices that evoke a sense of care that is highly situated and poly-contextual. At the same time, the experiences of such a practice-oriented research agenda opens up the possibility of getting a better grasp of general care issues about the embodied human social being (cf. also Latimer 2000; Latimer and Schillmeier 2009; Schillmeier and Domènech 2009). *New Technologies and Emerging Spaces of Care* unfolds novel ways of analyzing and reflecting upon the enactment of human society and how it is cared for. Social studies are studies of care; and care studies make us aware of the multiple orderings and disorderings, stabilities and instabilities, knowledge and ignorance, sameness and otherness of the embodied human society. Only by avoiding the 'fallacy of misplaced concreteness' (Whitehead 1967), that is, by taking care *not* to conflate the specific with the abstract and by avoiding explaining the former with the latter, *concepts of care* may help to take care of the specificities of human social life. In this kind of reading, social scientific research and theoretical reflection is about taking care of human society, which remains to be explained by the different and heterogeneous ways humans and non-humans relate or do not relate. The relationship between new technologies and emerging care spaces articulates a most important field of research. It provides insights, questions and controversies about how we live *and* how we would like to live.

Guiding the Reader

The book is subdivided into two main sections that seek to address these relationships. In section 1, the contributors address how new technologies enact new forms of care that question and alter non-institutional types of care. According to the dominant (but often idealized and ethnocentric) concepts, 'the home' is considered the place for care. Supposedly, the stable relationships that characterize this place make it a privileged locus for continuous care, in such a way that interdependence and independence appear as intermingled processes (Mallet 2004). This is one of the main reasons mentioned, leaving aside economic and political pressures, when giving meaning to the increasing introduction of health technologies in the home. Until very recently, chronic disease, disability or illness associated with aging used to have highly institutionalized solutions and treatments. The elderly and disabled often had to move from their private houses into long-term nursing homes. Frequently, such a move indicates a crucial incision into the commonalities of people's everyday lives, relationships, practices and emotions (cf. Challis et al. 2009; Fischer 1991; Gubrium and Sankar 1990; Milligan 2009; Schillmeier and Heinlein 2009).

Since the development of telecare and telehealth technologies, the home is recovering its central role in care practices (Blythe, Monk and Doughty 2005; Outshoorn 2008). Thanks to new technological and scientific innovations novel types of care practice have been introduced enabling increasingly frail people to remain in their own homes for longer. The use of information and communication technologies and the increased emphasis on robotics in solving certain problems open new possible interactions between the elderly and disabled and their different ways of living at home (Sparrow and Sparrow 2006). Through the use of knowledge and experience with new technologies, the sense of place has been altered and provokes a profound shift in people's sense and experience of their homes and affects complex reconfigurations of how people live and experience their personal and social concerns of everyday life.

Technological innovations also configure and reconfigure institutionalized spaces of care. Hospitals, long-care facilities or care-flats and traditional care places for elderly or disabled people are being confronted with profound transformations linked to the introduction of new care technologies. Thus, in section 2, the role and impact of (assistive) technology for elderly and (bodily or mentally) disabled people in relation to concepts and care practice spaces will be analyzed from international perspectives.

If in the first section we outline how homes are being institutionalized, in the second we explore how traditional institutions are losing most of their traditional and common traits in order to be more home like. The second section, then, analyzes in particular the role of technology and spatial rearrangements in the production of new forms of care, as well as highlighting the ideals and values that are realized and materialized in these new care regimes.

Section 1. Care at home

The collection of chapters presented in the first section highlights and compares some of the most relevant issues addressing the shift in the meaning and experience of the home in various national settings.

Christine Milligan, Maggie Mort and *Celia Roberts* are interested in moving topologies of technology and care. Given the current and projected growth of those in the older age groups and an ideological shift toward 'ageing in place', many of the technologies are targeted towards supporting the care needs (or perceived needs) of frail older people within the home. This 'technological fix', then, opens up new possibilities for enabling increasingly frail older people to remain in their own homes longer. Care providers suggest, that with careful design, the everyday presence of these technologies can be rendered 'invisible' as they become normalized within everyday routines and relationships between people and things. Although these are laudable goals, Milligan, Mort and Roberts argue, that neither the implementation nor the impact of these technologies should be accepted uncritically. Rather, these new forms of care have the power to reshape not only the physical and affective nature of home, but also the older people's experiences and relationships *within* the home. These themes are explored through an interrogation of empirical fieldwork conducted in the UK. Underpinned by debates about the meaning of home in relation to self and identity, the conceptualization of home as 'haven' (a place of refuge), and the home as a preconscious sense of setting. They suggest that the development and implementation of new care technologies may be reconfiguring both the experience of home and the places through which care relationships occur, pointing to the emergence of a new topology of care.

In his chapter, *Daniel López* offers a critical diagnosis of the care-delivering model that telecare technologies are bringing about. The materiality and spatiality of care and processes of subjectivation involved are discussed. Drawing on ethnographic material of home telecare services, López links his findings with insights from Surveillance Studies and the Foucauldian concept of governmentality.

The outcome is a map of intelligibility that enables us to understand the differences and continuities between telecare and traditional care-giving institutions. According to López, although clearly connected to ethical issues and power, the contribution of surveillance studies or governmental studies to a critical analysis of telecare developments has been limited so far. He believes that pointing out the post-panoptic features of telecare should not lead us to ignore those characteristics of telecare that are closely linked to traditional care settings such as, e.g. corporal discipline. We should not forget, López reminds us, that telecare requires the user to be disciplined. A panoptic society, according to López, is not a society configured by confined spaces, but a process through which normalizing surveillance is socially diffused by different micro-technologies of power.

Understood in this way, telecare is a kind of panopticon that manages the life of the users at home and whose most salient trait is the never-ending production of inscriptions gathering information about them (what condition or disease patients have, what pills they take etc). This provokes a new logic that goes beyond the discussion about the (post-) panopticon regime. The role of such inscriptions has to do with the need to reduce the risks associated with living independently at home. Foucault defines this risk management as security, and this is what enables a social diagram to be drawn that connects telecare with other contemporary systems. The idea of security enables López to rethink telecare and other technological solutions of community care as a contemporary tendency to securitize care spaces. For López, the difference between being at home or at an institution looses importance because any context, whether enclosed or open, is configured as a security space.

In the next chapter, *Chris Tweed* shows how the use of information and communication technologies to allow older people to stay in their own homes rather than in hospital or move to residential care facilities creates many new possible interactions between occupants, the building, and new devices. Usually, there are person-centred methods of assessing needs and risks that take into consideration the physical and psychological health of the patient as well as social context. Less emphasis is given to the impact telecare may have on the way people experience their homes or the changes it may bring to spatial practices within the home. The author examines existing theories about how people experience and use space and considers how these might stipulate innovative studies concerning the impact of telecare technologies and systems on spatial relations. Drawing on phenomenology and ecological psychology, Tweed considers two main concepts to be most relevant: the phenomenology of 'lived space' and J.J. Gibson's concept of 'affordances'.

Merleau-Ponty's phenomenology is taken as a useful starting point emphasizing the body as the primary locus of being-in-the-world (cf. Merleau-Ponty 1962). For the most part, though, Merleau-Ponty's bodies are healthy, full of vigour and even athletic, far removed from the experiences of the bodies of the elderly who might require increasing levels of care as their physical capabilities decline. Thus, while Merleau-Ponty offers valuable insights into the relations between bodies, space and perception, they need to be extended to account for people with a wide range of abilities *and* disabilities.

To do so, Tweed links Merleau-Ponty's phenomenology with J.J. Gibson's theory of affordances. This approach helps to suggest new ways of investigating how telecare-related technologies construct new affordances within the home. The author analyses the potential of the concept of affordance to inform the design of future telecare systems for people to use in their own homes drawing on his previous study of technologies in the home aimed at supporting independent living. Tweed concludes that space is not uniform or homogenous. Consequently, a Cartesian description of space is rather limited when we consider how people experience *lived space*.

In his chapter, *Peter A. Lutz* introduces the notion of *clutter moves* as an experimental heuristic for tracing how movement threads together a range of cluttered entities in ageing home care ecologies. In particular, he is concerned with older people and their cluttered technologies. The empirical data originates from an ethnographic study of ageing home care in the United States. According to his analysis, the category of *home clutter* is revealed as more complex than it is commonly portrayed in popular or scholarly accounts. The literature frequently cites general household clutter as a hazard in the domestic environment. In such reports moving around between cluttered things tends to provoke a dangerous outcome for older people, but removing home clutter turns out not always to be easy or appreciated among older people, for whom cluttered things are often revealed as meaningful and ordered.

Moreover, the movements of people and home clutter are highly entangled with home ecologies of ageing care. *Clutter moves*, then, is a powerful way to rethink the relationship between older people, their clutter and ageing home care ecologies including its technologies. The author suggests that when moves between collecting and distributing clutter are considered, it is possible to appreciate how home clutter may enable people with reduced mobility to have additional movements. The chapter ends with a conceptual shift from the standard view of technological clutter to a more complex view of *clutter technology* to extend the heuristic of *clutter moves*. This then is used for rethinking new technologies in relation to ageing home care ecologies, problematizing the very idea of the 'new': what qualifies as *new* and in relation to what? As the author notes, any degree of *newness* is largely a matter of perspective and framing. In a future landscape where home is increasingly equipped with healthcare devices and transformed into care facilities in their own right, the socio–technical notion of clutter technology emerges as an interesting tool for the further exploration of the implied transformations.

Hanne Lindegaard and *Søsser Brodersen*'s chapter explores how assistive technologies are shaped within everyday life in private dwellings. The authors draw on empirical material from an ethnographic research project carried out in Denmark, 'Homespace or Workspace?', which allows them to analyze the socio-material spaces in the use and distribution of assistive technologies. Given that, in Denmark, disabled and elderly people are able to receive domiciliary care from the municipality, they consider 'home' as becoming a hybrid of workspace and private space due the introduction of health workers and several assistive technologies. Lindegaard and Brodersen underline that the continual optimization of professional health care practice involving many assistive technologies often collides with the disabled person's expectations in relation to the kind of non-human actors they prefer in their homes. Obviously, whenever their expectations are not met by these new technologies, they frequently feel that being at home appears highly institutionalized and un-homely.

Accordingly, Lindegaard and Brodersen pose several questions concerning the hybrid nature of such spaces. How does everyday practice change in the interplay with institutionalized artefacts and other actors in domiciliary care? How do disabled people integrate the technologies or reject them? Is it possible to make them 'fit' into both everyday practices and their homes? To answer these questions Lindegaard and Brodersen mobilize several resources comprising work on socio-material relations with a focus on Actor-Network-Theory (ANT) and Domestication Theory. They explore how and when the actors and the various artefacts are able to become an aligned network or *when* and *why* these relations may fail. At the same time, Lindegaard and Brodersen pay attention to the aspects of the very meaning of artefacts, of how the symbolic relates to practice.

They focus on the complexities of so-called patient lifters (cf. also Schillmeier and Heinlein 2009). As with many other assistive technologies, the patient lifter seems to be designed, not only to assist the disabled, but also to meet the requirements for health care practice and assist the nurses. As a result, although many patients are 'lifter users' in their everyday lives, the lifter is not domesticated as an incorporated artefact. Hence, the relation between the artefact and the patient is not an aligned, ready-to-hand network that underlines the very feeling of being at home. Lindegaard and Brodersen conclude that to design assistive technologies, designers, care providers and politicians should acknowledge more explicitly the complexity with which the assistive technologies are to be applied and used, and come to a better understanding of for whom, for what and where the assistive technologies are, to achieve good care practice.

Robots have left the military and the factories and are being invited into peoples' homes. In the final chapter of Section 1, *Mark Paterson* analyzes the interactions between humans and robots, which are becoming increasingly ubiquitous. According to Paterson, there are several reasons that explain this phenomenon: a profit-motivated robotics industry inevitably seeking diversification, post-industrial, high-tech, high-density nations with increasingly aging populations and coincidentally high levels of consumer spending, economies sustained in the service sector and a general emphasis on knowledge transfer from higher education research institutions. Robots, as it seems to Paterson, are going to be everywhere. As robots are diversifying in their nature, reach and abilities, it is reasonable to think that the potential client groups are similarly diversifying and new sectors of society will benefit from robotics. Thus, mobility impaired people and the elderly could obtain physical assistance, rehabilitation and other therapeutic advantages.

Given such a future scenario, Paterson wonders what kind of social presence robots currently have, and what kind we want them to have. To answer this, he looks at robotics and human interaction through a series of case studies, and structures the chapter according to themes pertinent to the social presence of robots in assisted living contexts. He begins by looking at a grounded, mundane example of an available technology originated in a military-industrial context, but which is currently receiving much interest from the UK government for future domestic use: a remote-controlled 'snake-arm' robot designed for use within confined spaces. From this existing example and its context the author looks at the purpose and policy of service robots in general, examining and comparing strategies and approaches in Europe, the US and Japan. The conclusion is that the same potential demographic problem is being approached in different ways. These ways reveal more of the cultural-historical construction and reception of robotics in these geographically disparate locales. According to Paterson, healthcare institutions and care homes could certainly benefit from the aid of robots: for physical assistance, emotional presence or drug delivery, to name a few. Given the diversification of client groups due to demographic shifts such as the increased population of the elderly and the relative paucity of immigrants to care for them, along with the reduction in manufacturing costs, the author envisages an increasing number of relatively low-cost robots in the home. This leads Paterson to consider the mixed and sometimes confined spaces of physical human-robot interactions (pHRI). Several questions arise: how are non-traditional client groups to negotiate with robots' everyday spatial presence? How will this occur in domestic spaces, and what is the nature of the resulting mixed or hybrid spaces of human-robot interaction? To answer these questions the author follows two concepts extrapolated from their deployment in human-computer interaction (HCI) to human-robot interaction (HRI): presence and proximity. Robot interactions occur in hybrid human-machine spaces, depending on the human perception of a non-human presence that must become 'social' or even 'sociable' in order to seem more naturalistic. Thus, sociable robots come equipped with the very abilities that humans have evolved to ease our interactions with one another: eye contact, gaze direction, turn-taking or shared attention. Nevertheless, this is not as straightforward as it might seem. Indeed, in order for non-traditional client groups to willingly accept physical assistance from robots, the unfamiliarly machinic and non-humanoid may not appear as uncanny.

Section 2. Re-forming care homes

The opening chapter provides insights and provokes reflection concerning dementia care. *Hilde Thygesen* and *Ingunn Moser*'s contribution is structured around three main questions that they try to answer: What constitutes good dementia care? How is this achieved? And, what is the role of technology therein? To do this, they draw on empirical data from ethnographic observations of dementia care practices, in four different care homes, for people with dementia in the Norwegian context. Their approach is situated within the tradition that highlights ethics as an empirically unfolding issue and not as an abstract, normative and given objective. This means that the question concerning what constitutes *good* dementia care is not answered from the outside, according to different general ideas of what good care should be. Rather it emerges within the situated ideals, limits and reflections in and of care practice. This empirical ethics is combined with a semiotic tradition that treats subjectivity as relational, but extending the relations further than the discursive realm and including the material in the widest sense.

Examining accounts of 'fall prevention', Thygesen and Moser explore how various technologies, and, in a broader sense, how material arrangements are mobilized in the continuous, demanding and creative process that constitutes everyday care. This leads them to affirm that good dementia care may be understood as an ongoing process of trying out and assessing different care arrangements. This process is considered ethical in the sense that it involves a number of different values that are related, weighed and balanced in practice by carers. Concluding, Thygesen and Moser consider that good dementia care has to do with what they call 'sustaining the person'. 'Sustaining the person' refers to those care practices that enable the person to achieve, develop and/or maintain a sense of self, of an 'I', understood as a position that also has attributed subjectivity or personhood by others. By employing the notion of 'sustaining the person' Thygesen and Moser move away from an understanding of care practices as dependent on the idea of a person as autonomous, rational and independent in order to allow for multiple and contrasting values to co-exist in the enactment of what a person is and what makes a person. Furthermore, the notion of 'sustaining persons' implies a collective rather than an individualized focus. Care is not just about meeting individual needs, but involves a large network of humans and technologies involved in shared daily life and in the caring process. Emphasizing the situatedness of care and its ethics, Thygesen and Moser are challenging current policy on technology and care, which is based on an understanding of good care as a given entity.

Elena Bendien, Steven D. Brown and *Paula Reavey* invite the reader to look at a very peculiar care home in the Netherlands. This care home is ruled by a care-provider organization specialized in care for the elderly and is conceived as a place of continuous activity where elderly clients have continuous opportunities to focus their attention on things beyond their own immediate health. As the authors show in their analysis, the home care constitutes in itself an alternative version of care to that of biomedical approaches. Thus, it is a strong commitment of the organization to emphasize the social relations between staff and clients as the primary site where care occurs. Nevertheless, unlike a classical model of care, where the client is dependent on the expertise and relative power of the care-provider, here, this relationship is inverted and care often happens precisely when the clinician refuses to automatically exercise expertise. Following the principle that 'too much care is as bad as too little care', caregivers are often reminded to facilitate self-care as far as possible. Bendien et al. pay special attention to one of the facilities of the care home: the reminiscence museum, where an ethnographic study was conducted over eighteen months. This museum consists of a series of specially constructed rooms built in the basement of the house in such a way that they resemble a Dutch domestic space from the first half of the twentieth century. Bendien et al. analyze the museum as a very particular kind of care setting that redefines what is meant by 'care'. Indeed one of the most interesting features of the museum is that it is specifically intended not to be a care setting at all. Its purpose is simply to facilitate 'happiness'. It has nothing to do with care in the formal medical sense of the term. Actually, the care organization sees the museum as being designed to support an 'art of living', which takes the notion of living as the process of giving meaning to existence, as an ongoing questioning of what it is to be living this particular life.

Finally, adopting a contemporary STS approach, Bendien et al. regard the reminiscence museum as a 'social technology'. That is, a form of technology that not only has a relatively immaterial basis in their enactment, but which is directed at transforming social relations. Thus, thanks to the joint construction of memories about past times that occurs through the interaction of visitors, staff and the collected objects, the museum becomes a structured social space for the enactment of practices of remembering. This work of remembering espouses no therapeutic goals. It is not confined to reproducing or more-or-less accurately recalling the past, nor is there encouragement for visitors to view the past as a coherent story that defines their present circumstances. Instead, the visit provides the opportunity for a relative redefinition of the present through the mobilization of the past. As Bendien et al. conclude, 'the art of living consists not of transforming the present into a new future, but rather of opening up the past from its apparent rigidity and determination to afford new possibilities in the present' (p. 166).

Kraeftner, Kroell, Ramsebner, Peschta and *Warner* of *Xperiment*, an Austrian transdisciplinary research group, use the relationship with pillows, maintained by severely disabled patients needing long-term care, as an opportunity to sensitize readers to the intricate relationships between humans and artefacts. To do this, firstly, Kraeftner et al. describe various assistive and conventional technologies that configure the dwelling of patients diagnosed as persons in a vegetative state. This condition is pre-dominantly caused by a severe brain injury, and, according to the standard definition, it means that these patients do not have any reactions to their environment or by themselves. Kraftner et al. are interested in technologies that are meant to change the surroundings and the respective dwelling places of patients. Usually the goal is to 'reshape' their bodies and avoid the consequences of their inert condition.

Kraeftner et al. underline that almost none of these care practices, be they more or less technology driven, remain uncontested but rather give rise to debates. Secondly, they discuss several technically heterogeneous diagnostic techniques that quite recently started to modify the definitions of the syndrome, turning it into a very complex affair. Third, they discuss the different logics of care that may come with the implementation of different technological approaches. What is interesting in their approach is a shift from epistemological questions to ontological ones. This leads them to avoid asking about how to diagnose, treat or care according a pre-given idea of syndrome. Rather, attention is paid to the different versions of syndromes as they are enacted along technological developments, practices of diagnosing, treating and caring. In this vein, special focus is given to multiple articulations of artefacts and humans, and not determined pre-given beings and conditions. Fourth, Kraeftner et al. make a historical detour focusing on the vicissitudes of the lie detector, also known as the polygraph that inspires the authors with the possibility of constructing a combined sensor: a technical-human hybrid that would be able to detect whether patients in vegetative states are minimally conscious. They conclude this section by making a risky wager: their clinical 'research in the wild'/'pillow research' will be able to detect awareness and behaviours in 'non-behavioural' patients in a minimally conscious state. Fifth, they explain how they develop a proprietary imaging technology, a so-called technical-human hybrid sensor. This sensor would be able to produce a 'diagnostic set-up' with four main characteristics: retaining the 'thickness' of clinical context, accepting 'formal/informal' and expert/non-expert components of information, standardizing the presentation of emotional stimuli, and, finding out what kind of stimuli the patient prefers.

Summing up, through *Pillow Research* and the prototypical development of an alternative imaging technology Kraeftner et al. intend to enrich the discussions about the clinical care of severely disabled people. Instead of relying on the well-known classical dichotomies such as expert/non-expert, patient/non-patient or human/technical, their 'participatory interventions' seek to create problematized standard visions of care and to produce concern and interest capable of enrolling actors and assigning them new and surprising roles.

Alexandra Hillman, Joanna Latimer and *Paul White* round up the last chapter of Part II. They examine how access to care is accomplished in medical settings by drawing upon ethnographic work in a UK regional teaching hospital (University Hospital) of three settings in clinical medicine: Accident and Emergency, Medical Genetics and Intensive Care. In particular, Hillman et al. emphasize the role of managerial and clinical technologies in the process of gaining access to care. As they argue, gaining access to health care is far from being straightforward, rather it is itself increasingly managed by producing particular associations and disassociations, attachments and detachments in the day-to-day organization of clinical settings that demand the exclusion and inclusion of people, work and things. Through boundary work and the multiple means through which they are bodied forth, tacit understandings of the technology involved make visible the relations between legitimate patients and the work of institutions in (re)producing themselves as particular kinds of medical practices.

Each analysis of the three settings is interesting in itself. Thus, the examination of the Accident and Emergency Service leads the authors to show how the mediation of technology for administering, rationing and auditing emergency medicine operates as a regulating device of the behaviour of those that come into contact with it. Hillman et al. observe the infiltration of discourses relating to efficiency and resource management with the process of clinical decision-making. One effect of this is expressed in a shift of relations between staff and patients, particularly in assessment processes whereby patients are pushed to 'sort themselves out' according to their own responsibilities as health care users. In the analysis of the Intensive Care Unit, Hillman et al. offer an example of complex ways through which managerial policy communicates the relations of intensive care staff with patients. In this case, they show the ways in which staff's appropriation, alignment or resistance to the changing policies for admission procedures helps them to (re)draw the boundaries of Intensive Care as a mode of ordering itself and others. As a consequence of these moves particular care spaces emerge.

Contrasting with ordinary analysis devoted to medical technologies and their impact upon care, Hillman et al. try to reveal managerial technologies and their effects on people. Finally, Hillman et al. carry out an analysis of a service devoted to medical genetics. In this case, we are confronted with the reverse side of the same issue regarding access. The problem is no longer that of keeping people out, but keeping them and relevant legitimating practices on. We can see how the geneticists' interest is not really in 'individuals', but the extended family across, and up and down the generations. What emerges in the analysis is that in the genetic clinic the work of 'keeping people on' is as important as the work of making a diagnosis, particularly because arriving at a diagnosis is not an easy or unproblematic affair. In the end, it is the parents' and sometimes even the grandparents' capacity for some kind of commitment that makes the difference.

Hillman et al. offer a beautiful closing chapter providing a lucid reflection on how different kinds of technologies figure within the associations and orderings of everyday health care work by examining what they accomplished in terms of specific forms of orderings and processes of inclusion and exclusion.

References

Bloomfield, B. P. and McLean, C. 2003. Beyond the Walls of the Asylum: Information and Organization in the Provision of Community Mental Health Services. *Information and Organization*, 13, 53–84.

Blythe, M. A., Monk, A. F. and Doughty, K. 2005. Socially Dependable Design: The Challenge of Ageing Populations for HCI. *Interacting with Computers*, 17, 672–689.

Carrillo, M. C., Dishman, E. and Plowman, T. 2009. Everyday Technologies for Alzheimer's Disease Care: Research Findings, Directions, and Challenges. *Alzheimer's & Dementia*, 5, 479–488.

Challis, D., Sutcliffe, C., Hughes, J., von Abendorff, R., Brown, P. and Chesterman, J. 2009. *Supporting People with Dementia at Home*. Farnham: Ashgate.

Chambers, M. and Connor, S. L. 2002. User-friendly Technology to Help Family Carers Cope. *Journal of Advanced Nursing*, 40(5), 568–577.

Dyck, I., Kontos, P., Angus, J., McKeever, P. 2005. The Home as a Site for Long-term Care: Meanings and Management of Bodies and Spaces. *Health & Place*, 11, 173–185

Essén, A. 2008. The Two Facets of Electronic Care Surveillance: An Exploration of the Views of Older People Who Leave with Monitoring Devices. *Social Science & Medicine*, 67, 128–136.

Fex, A., Ek, A-C and Söderhamn, O. 2009. Self-care Among Persons Using Advanced Medical Technology at Home. *Journal of Clinical Nursing*, 18, 2809–2817.

Fisher, B. J. 1991. *It's Not Quite Like Home: Illness Career Descent and the Stigma of Living at a Mulitilevel Care Retirement Facility*. New York: Garland.

Gubrium, F and Sankar, A. (eds) 1990. *The Home Care Experience: Ethnography and Policy*. London: Sage.

Habermas, J. 1985. The Theory of Communicative Action. Boston: Beacon Press.

Harper, R. H. R. (ed.). 2006. *Inside the Smart Home*. London: Springer.

Hjalmarsson, M. 2009. New Technology in Home Help Services – A Tool for Support or an Instrument of Subordination? *Gender, Work and Organization*, 16(3), 368–384.

Latimer, J. 2009. *The Conduct of Care. Understanding Nursing Practise.* Oxford: Blackwell.

Latimer, J. and Schillmeier, M. (eds) 2009. *Un/knowing Bodies*. Oxford: Blackwell.

Latour, B. 1992. Where Are the Missing Masses? The Sociology of a Few Mundane Artifacts, in *Shaping Technology/Building Society. Studies of Sociotechnical Change*, edited by W. Bijker and J. Law. Cambridge: MIT, 225–258.

Latour, B. 2005. *Reassembling the Social. An Introduction to Actor-Network-Theory*. Oxford: Oxford University.

Laviolette, P. and Hanson, J. 2007. Home Is Where The Heart Stopped: Panopticism, Chronic Disease, and the Domestication of Assistive Technology. *Home Cultures*, 4, (1), 25–44.

Loe, M. 2010. Doing it My Way: Old Women, Technology and Wellbeing. *Sociology of Health & Illness*, 32(2), 319–334.

López, D. and Domènech, M. 2009. Embodying Autonomy in a Home Telecare Service. In *Un/knowing Bodies*, edited by J. Latimer and M. Schillmeier. Oxford: Blackwell, 181–195.

Magnusson, L., Hanson, E. 2003. Ethical Issues Arising from a Research, Technology and Development Project to Support Frail Older People and their Family Carers at Home. *Health and Social Care in the Community*, 11(5), 431–439.

Magnusson, L., Hanson, E. and Nolan, M. 2005. The Impact of Information and Communication Technology on Family Carers of Older People and Professionals in Sweden. *Ageing & Society*, 25, 693–713.

Magnusson, L., Hanson, E., Brito, L., Berthold, H., Chambers, M. and Daly, T. 2002. Supporting family carers through the use of information communication technology – The EU project ACTION. *International Journal of Nursing Studies*, 39, 369–381.

Mallet, S. 2004. Understanding Home: A Critical Review of the literature. *The Sociological Review*, 52(1), 62–89.

Mann, W. C. 2001. The Potential of Technology to Ease the Care Provider's Burden. *Generations*, 25(1), 44–48.

Marshall, M. 1999. Technology to Help People with Dementia Remain in their Own Homes. *Generations*, 23(3), 85–87

May, C., Finch, T., Mair, F. and Mort, M. 2005. Towards a Wireless Patient: Chronich Illness, Scarce Care and Technological Innovation in the United Kingdom. *Social Science & Medicine*, 61, 1485–1494.

Milligan, Ch. 2001. *Geographies of Care. Space, Place and the Voluntary Sector*. Farnham: Ashgate.

Milligan, Ch. 2009. *There Is No Place like Home: Place and Care in an Ageing Society*. Farnham: Ashgate.

Mol, A. 2008. *The Logic of Care. Health and the Problem of Patient Choice*. London: Routledge.

Mort, M., May, C. R., and Williams, T. 2003. Remote Doctors and Absent Patients: Acting at a Distance in Telemedicine? *Science, Technology and Human Values*, 28, 274–295.

Nicolini, D. 2006. The Work to Make Telemedicine Work: A Social and Articulative View. *Social Science & Medicine*, 62, 2754–2767.

Outshoorn, J. 2008. The Provision of Home Care as a Policy Problem. *Journal of Comparative Policy Analysis*, 10(1), 7–27.

Percival, J. and Hanson, J. 2006. Big Brother or Brave New World? Telecare and its Implications for Older People's Independence and Social Inclusion. *Critical Social Policy*, 26, 888–909.

Pickard, S. 2009. Governing Old Age: The 'Case Managed' Older Person. *Sociology*, 43(1), 67–84.

Poland, B., Lehoux, P., Holmes, D. and Andrews, G. 2005. How Place Matters: Unpacking Technology and Power in Health and Social Care. *Health and Social Care in the Community*, 13(2), 170–180

Pols, J. and Moser, I. 2009. Cold Technologies versus Warm Care? On Affective and Social Relations with and through Care Technologies. *ALTER European Journal of Disability Research*, 3, 159–178.

Roberts, C. and Mort, M. 2009. Reshaping what Counts as Care: Older People, Work and New Technologies. *ALTER, European Journal of Disability Research*, 3, 138–158.

Savenstedt, S., Sandman, P. O. and Zingmark, K. 2006. The Duality in Using Information and Communication Technology in Elder Care. *Journal of Advanced Nursing*, 56(1), 17–25.

Schillmeier, M. 2009. Actor-Networks of Dementia, in *Un/knowing Bodies*, edited by M. Schillmeier and J. Latimer. Oxford: Blackwell, 141–158.

Schillmeier, M. 2010. *Rethinking Disability. Bodies, Senses and Things*. London, New York: Routledge.

Schillmeier, M. and Domènech, M. (eds) 2009. Care and the Art of Dwelling: Bodies, Technologies, and Home. Special Issue *Space and Culture*, 12(3).

Schillmeier, M. and Heinlein, M. 2009. From House to Nursing Home and the (Un-)Canniness of Being at Home. *Space and Culture*, 12(2), 218–231.

Serres, M. 1994. *Atlas*. Paris: Editions Julliard.

Serres, M. 1995. *The Natural Contract*. [Translated by E. MacAuthur and W. Paulson], Ann Arbor: University of Michigan.

Shine, K. I. 2004. Technology and Health. *Technology in Society*, 26, 137–148.

Soar, J. and Seo, Y. 2007. Health and Aged Care Enabled by Information Technology. *Ann. N.Y. Acad. Sci.*, 1114, 154–161.

Sparrow, R. and Sparrow, L. 2006. In the Hands of Machines? The Future of Aged Care. *Mind Mac*, 16, 141–161.

Strum, S. and Latour, B. 1999. Redefining the Social Link: from Baboons to Humans, in *The Social Shaping of Technology*, edited by D. Mackenzie and J. Wajcman. Buckingham: Open University, 116–124.

Sybylla, R. 2001. Hearing Whose Voice? The Ethics of Care and the Practices of Liberty: a Critique. *Economy and Society*, 30(1), 66–84.

Van Berlo, A. 2002. Smart Home technology: have older people paved away? *Gerontechnology*, 2, 77–87.

Vaz, P. and Bruno, F. (2003). Types of Self-Surveillance: from Abnormality to Individuals 'at Risk'. *Surveillance and Society*, 1(3), 272–291.

Verkerk, M. 2001. The Care Perspective and Autonomy *Medicine, Health Care and Philosophy*, 4, 289–294.

Whitehead, A. N. (1967[1925]) *Science in the Modern World. Lowell Lectures 1925*. New York: The Free Press.

Williams, G., Doughty, K. and Bradley, D. A. 2000. Safety and Risk Issues in Using Telecare. *Journal of Telemedicine and Telecare*, 6, 249–262.

Chapter 1

Cracks in the Door? Technology and the Shifting Topology of Care

Christine Milligan, Maggie Mort and Celia Roberts

Introduction

The rapid growth of new and emerging care technologies (NCTs) targeted at supporting the care needs (or perceived needs) of frail older people within the home has been accompanied by an ideological shift toward 'ageing in place'. That is, a widespread belief in many advanced capitalist countries that the family (or individual) home is both the preferred and most cost-effective site of care and support for older people (Tucker et al. 2008; López and Sánchez-Criado 2009). Ageing in environments to which they are accustomed, is believed to enable older people to remain more independent, despite growing frailty. The supportive and enabling environment of the familial home is framed in sharp contrast to images of the stresses experienced by older people and their families as they find themselves having to move to residential care settings (Milligan 2006).

Ageing in place assumes that the process of inhabiting a place over time somehow results in the development of a distinctive sense of attachment and 'autobiographical insideness' (deriving from who a person was and who they have become) (Rowles 1993, 69) as the home comes to symbolize a person's identity. This sense of place and place attachment is viewed as being adaptive and supportive for older people. In policy terms this means implementing changes in housing and healthcare practices in order to ensure that older people have access to those technologies, products and services that will enable them to remain in their own homes as their needs change. Inevitably such ideological and policy shifts are turning the home into a reconfigured place – one that opens up new debates about the meaning and experience of home for older people.

Ageing in place combined with the current and projected growth of those in the older age groups and increasing concern about a potential 'care crisis' as demand for care exceeds provision, means that any turn towards a 'technological fix' opens up some new and potentially challenging possibilities for redressing the so-called 'care gap' (Lansley et al. 2004). At the same time, these dilemmas raise important questions about how older people actually experience new care technologies and the extent to which they are reshaping both the nature of care and the places within which that care is performed.

Neither the implementation, nor the effect of these technologies should be accepted uncritically. Indeed, commentators have pointed to the ways in which the reorganisation of the home to accommodate even relatively low level assistive technologies (such as hoists, lifts, rails, wheelchairs, commodes etc.) can reshape the nature and experience of home such that older people and their family carers no longer feel 'at home' (see for example, Milligan 2000; Wiles 2003; Angus et al. 2005). Yet providers and developers of NCTs maintain that with careful design, the everyday presence of these technologies can be rendered 'invisible' as they become normalised within the everyday routines and relationships that exist between people and things. We suggest, however, that whilst it may be possible to render some care technologies relatively invisible, the installation of technologies designed for surveillance and monitoring can create a fundamental shift in older people's sense and experience of home. NCTs thus have the potential to reshape not just the relationships between people and things that occur within the physical structure of the home, but also the feeling and sense of 'being at home' (Blunt and Dowling 2006). Any attempt to understand the effects of care technologies thus brings into focus the complexity of home as both a site of affect, social interaction and personal meaning, and as a site of care.

The purpose of the chapter then, is to explore the relationship between care, technology and home and the effects of these relationships on older people – specifically in England, UK (though many of the issues raised will be of wider resonance). Underpinned by debates about the meaning of home in relation to self and identity; the conceptualisation of home as 'haven' (a place of refuge); and the home as a preconscious sense of setting, we suggest that the development and implementation of NCTs may be reconfiguring the experience of home in new or different ways from that of earlier forms of adaptive technology. Whilst acknowledging the specifically western-centric nature of these debates, such a shift, we suggest, is giving rise to a new topology of care that has implications for successful ageing in place.[1] We illustrate our discussion by drawing on empirical data gathered from citizens' panels and ethnographic fieldwork conducted in England as part of a wider European study.[2] Before doing so, however, we briefly reflect on the sorts of care technologies to which we refer.

---

1   Successful ageing in place refers to a person's ability to reside in their place of choice (usually the domestic home) with access to services that meet their needs without having to move each time their needs change. It also includes opportunities to engage in meaningful social activities and community life.

2   The material has been collected as part of the EFORTT study (Ethical Frameworks for Telecare technologies for older people at home) funded by the EC Seventh Framework Programme, Science in Society call.

New Care Technologies – An Overview

Care technologies include a broad range of care support ranging from devices and systems that either enable individuals to perform tasks they would otherwise be unable to do, or increase the ease and safety with which these tasks can be performed (Cowan and Turner-Smith 1999). These technologies can be (and often are) used within domiciliary settings and include telecare systems and 'smart homes'. The former refers to technology designed to improve existing home-care services (Demiris et al. 2004), whilst the latter encapsulates technologies aimed at facilitating non-obtrusive risk management and monitoring within domestic space. Tinker et al. (2004) further distinguish between what they refer to as portable assistive technologies (such as alarms, monitors, motion detectors etc.) and fixed assistive technologies that may require housing adaptation (such as lifts and ramps).

Whilst acknowledging that a wide range of care technologies, including assistive devices such as hoists, canes and rails have been commonly available for many years, in this chapter we are concerned with newer and emerging care technologies (based on information and communication technologies) designed to both monitor the physical health and activity patterns of (primarily) older people and to support their ability to age in place. These include:

- environmental control solutions (e.g. wireless control for electronic equipment in the home that can also be linked to telecare solutions);
- remote telecare and diagnostic systems (e.g.using internet and webcam technology);
- electronic 'reminder systems' (i.e. devices designed to set off an alert call if an activity – such as taking tablets – is are not taken at appropriate times);
- wearable or 'smart home' devices that monitor and gather continuous data (e.g. motion and falls detectors, monitors that measure frequency of use of household appliances and facilities such as fridges, food cupboards, toilets etc.) and intruder alarms;
- 'smart clothing and fabrics' with inbuilt sensors to monitor health status such as heart rate, abnormal heart activity, pulse, temperature etc;
- electronic monitoring and tagging devices designed to identify patterns of movement and location.

The growth of these technologies is likely to be enhanced by people's increasing familiarity with wireless networks and mobile phones – facilitating the creation of a consumer platform for them. The growth of wireless networks not only offers the potential to replace more visible wired home technology, but being easier and more flexible to install, they are also less costly and (it is argued) cheaper to maintain (Tinker et al. 2004). As such, NCTs are seen as an increasingly attractive proposition for care providers. In the UK, for example, the Government announced an investment of some £80 million through a two year Preventative Technology

Grant (from April 2006–March 2008)[3] with the aim of initiating changes in the design and delivery of housing, health and social care. These technologies are not only seen as having the potential to enhance and maintain the well-being and independence of a wide range of individuals who would, arguably, be otherwise unable to live independently in the home, they are also seen as part of a strategy to reduce the number of older people entering residential care and hospitals (Bayer et al. 2007). The Government in England, for example, specifically stated its belief that over the period 2006–2008, the Preventative Technology Grant would reduce these numbers by some 160,000 older people (Department of Health 2008).

The Meaning of Home and Identity in the Care Relationship

The home is a setting that is both familiar and imbued with particular meaning. Whilst acknowledging that for a minority the home can represent a 'scarred space' – that is, a site of fear or abuse – for others, the ongoing temporal process of inhabiting a familiar place can result in the development of a unique sense of attachment that can be both supportive and adaptive – particularly for older people. There are at least three core aspects that contribute to this. We elaborate on these below.

Home, Self and Identity

Firstly, home can be viewed as a materialization and container of identity, emotion and self-expression – one that not only anchors people within a particular locality, but which is manifest as a site of memories and a daily reminder of continuity with past identities and relationships. Augé's (1995) notion of 'anthropological place' offers a way of facilitating our understanding of the importance of place and home in the care relationship. As a site of embodied memories and family history, the home has the potential to facilitate an understanding of an inhabitant's needs and desires. The presence of private possessions and familiar objects within the home can also act to reinforce a person's sense of self and social status, endowing the home with personal meaning (e.g. Rubinstein 1989; Davidson 2009). A place that is inhabited by the same person over a period of time thus comes to portray that person based on objects (present or absent), their relationship with those objects, the habits they imply and the emotions they incur. Exclusions and preferences, order and disorder, the organisation of available space, relationships formed with objects within those spaces, colour, texture and the use of materials all serve to compose a 'life narrative' about those living within that home (de Certeau et al. 1998). As such, the home provides important clues about the identity and

---

3   Grant funding ended in March 2008, although any remaining monies were allowed to be carried over into 2008–09. Local authorities now bear the responsibility of providing mainstreamed telecare services.

personality of the older person, their ability or inability to cope that can be critical in the construction and performance of effective caring relationships. As a site of materialized identity, the home also holds the potential to place limits on 'the extent to which an individual can be objectified and depersonalised as within a collective [institutional] regime' (Milligan 2009, 69).

Home as 'Haven'

People's attachment to home is also seen to be linked to their perception of home as a site of security, familiarity and nurture (Tuan 2004). That is, it represents a place to which the individual can retreat from the public gaze; where he or she can exert a level of control in relation to the performance of practices that take place within it; and where he or she holds a greater level of power in relation to decisions about who to admit or exclude. This can be particularly significant for older people who feel vulnerable outside the bounds of their own private space. Rowles (1978) and others, for example clearly demonstrate how, with increasing age, older people's life-worlds and the spaces they inhabit contract. Indeed this was reinforced by participants in our own research:

> Elsie[4]: and I think another thing about friends is that as you get older… not ageing but I have actually lost I should think half a dozen friends who have died. The older you get the more this is happening.
> (agreement round the room)
> Bill: the more this is likely to happen
> Nancy: you're less mobile and you can't get out and your social circle, social contact disintegrates.

The home and immediate surrounds thus become increasingly important in helping to reinforce an older person's sense of self (Sacco and Nakhaie 2001). Familiarity with the organisation of the home and immediate surrounds can provide an important reinforcement to the older person's sense of self and safety. As a site of ontological security (Giddens 1990) the home also becomes a familiar and 'safe space' from the perceived threats and challenges of the outside world and so is critical to the older person's ability to successfully age in place. Here, ontological security is defined as the confidence that most human beings have in the continuity of their self-identity and in the constancy of their social and material environments (Giddens 1990). The home thus represents a site of constancy and a secure base around which to construct identities in the social and material environment.

Older people who have a positive attachment to home are thus seen as more likely to feel secure, in control of their lives and to have a more affirmative sense of self, facilitating 'adjustment to the contingencies of ageing and enhancing well-being' (Wiles et al. 2009, 665).

---

4   All participants' names have been replaced by pseudonyms.

Home as a Preconscious Sense of Setting

Rowles (1993, 66) suggests that, over time, people develop a 'preconscious sense of the setting' that reflects not only their physical and emotional attachment to the home, but also to the routines performed within it. An older person's preconscious sense of setting is thus seen to enhance his or her ability to negotiate spaces instinctively within the home without coming to harm. That is, despite increasing sensory or mobility impairments, the older person is able to negotiate rooms, doors, light switches etc. based on historical knowledge and memory of the home, its layout and the objects within it. However, whilst this preconscious sense of setting may enable individuals to transcend apparent physiological and sensory limitations, it may also make them more vulnerable to changes in that environment. A small object left on the floor or the moving of a small item of furniture (such as a chair or side table etc.) to a different location can, for example, disrupt the older person's preconscious sense of setting resulting in a fall or other accident. Larger changes in the physical environment arising as a consequence of installing adaptive aids, technological care supports and the wider paraphernalia of care, or rearranging the home to facilitate care work, are thus likely to interfere with the preconscious sense of setting and increase vulnerability. Given that sensors are placed on walls, smoke detectors on the ceiling, bed sensors under the mattress and so forth, NCTs could be argued to leave the physical environment largely unaltered. However, the central 'hub' which activates these devices is placed in a central position near the telephone and the user is usually expected to wear a device around his or her neck or wrist at all times. When activated, these systems give out audible alarms, beeps and voice messages, all of which hold the potential to disrupt an individual's preconscious sense of setting.

Telecare Technologies and Older People's Relationships with Home

The previous section illustrates how the temporal process of inhabiting the home can result in the development of a unique sense of attachment that can be both supportive and adaptive for older people. Such claims could underpin, and indeed support, the policy drive toward ageing in place. Yet as people age, not only are the spaces they inhabit likely to undergo physical transformation, but older people's sense of place and their associations with those places can also change (Wiles 2005; Schillmeier and Domenéch 2009). Hence the home cannot be seen as a static physical structure, but rather as a dynamic, negotiated and contested space in which understandings of home and the complex relationships performed within it, are played out (Milligan 2009). Of particular relevance here is the way in which the integration of new care technologies contributes to the emergence of a new topology of care – one that is reshaping care and care relationships both within and across the traditional boundaries of home. This then, raises critical questions about how the performance of care through and around these technologies is facilitating or disrupting older people's preconscious sense of setting and experiences of home, and what this may mean in terms of their ability to successfully age in place. In the remainder of this section, we consider these questions by focusing on four specific themes: monitoring and surveillance; ontological security; the changing nature of care interactions; and the shifting topology of care.

Monitoring, Surveillance and Home

Many current and emerging care technologies are designed to monitor older people's everyday routines within the home. In the smart home, sensors strategically located around the home collect daily activity data that is aggregated, tabulated to display patterns and then stored in databases. Any deviation from these pattterns (for example unusual movements during the night, multiple visits to the bathroom, failure to get out of bed, visit the kitchen or open the fridge) then triggers a signal alerting a 'distant other'. Falls monitors, smart fabrics and self-activated wearable devices are similarly designed to trigger an alarm in the case of an accident or adverse event. The 'distant other' (most commonly a call-centre operator) is then charged with assessing the situation and taking appropriate action – either contacting a named relative, friend or neighbour to check on the older person, or the emergency services if necessary. Where social networks are thin, call-centre operators may contact paid service responders.

Within our research, care professionals appear to make little or no distinction in how they view (the purpose of) wearable devices, fixed sensors and surveillance technologies – all of which come under the term 'telecare' and are subject to assessment of need/circumstances. Some older people however, view particular kinds of devices with considerably more suspicion and mistrust than others. Monitors and sensors are frequently referred as being 'big brotherish' and an invasion of the personal and private space of the home. Indeed, older citizens' panellists within our study maintained that, in their view, the installation of room sensors would significantly affect the meaning of, and attachment to, home:

> Karen: I wouldn't feel at home, I'd feel uncomfortable all the time.
> I: it wouldn't be like home any more?
> (Chorus of 'no')
> Nancy: you'd get frightened of moving around, thinking somebody's watching you.
> and:
> Betty: I wouldn't like somebody to know every time I went out through the front door.
> 2 women say: No, exactly
> Betty: it's very intrusive

In reflecting on the findings emerging from their own work, Magnusson and Hanson (2003) go so far as to claim that the intrusion of surveillance care technologies into the home represents not only a breaching of privacy but a potentially 'unjustified paternalism'; hence they argue that such technologies need to be treated with extreme caution.

Important questions about privacy were also raised in relation to consent and who has access to the activity data. In particular respondents were concerned about systems that held the potential for adult sons and daughters to access data revealing their everyday movements and routines. As one respondent put it:

> Frank: it's bad enough having them [adult children] ring up on the phone ...
> 'where were you yesterday?' you know – 'we rang earlier and you weren't in'.
> The thought that we... ooh

This highlights one of the dichotomies of care that arose in our study: that is, older people's need and desire to retain control over their own lives for as long as possible versus the fears and concerns of formal and informal (family) carers for their safety and an emphasis on risk avoidance. This reinforces work by Tracy et al. (2004) in the Canadian context that illustrates how family carers like to be kept informed of the care-recipient's condition even if this comes without the latter's consent. Further, they note that while care professionals often value the disclosure of information to colleagues and informal carers – justified as being 'in the patients' best interests' – this view is often at odds with older people's desire to retain control over disclosure to informal carers and other family members. Indeed, Morris (2005) notes that the disclosure of information to family carers can not only affect relationships with the ageing parent (who may feel aggrieved at having his or her privacy invaded), but also those between siblings, where contention arises over who is, or is not, participating in the care of the elderly parent. Surveillance technologies in the home thus hold the potential to substantially alter family relationships.

Concern about surveillance data is not solely focussed on third-party access to electronically stored medical information; it also involves issues of psychological privacy (the revealing of intimate attitudes, beliefs and feelings); social privacy (the ability to control social contacts); and physical privacy (the ability to control physical accessibility) (Miller 2001). Others, however, argue that such critical interpretations of the effects of surveilling care technologies present a one-sided dystopic view – one that they claim is both superficial and analytically unfounded (see for example, Lianos 2003; Blythe et al. 2005; Essén 2008). Indeed, such commentators often support their claims by arguing that levels of surveillance within alternative sites of institutional care are significantly greater – a tautological debate that rests on the degree rather than the presence or absence of surveillance within differing care settings. Furthermore, as the following interview excerpt with a representative from a care technology manufacturer illustrates, future developments in home-based NCTs hold significantly greater surveillance potential:

> Derek: I think the development of this type of technology over this next decade
> will move very, very quickly, and it will change beyond recognition. There will

be new technologies that come through that supersede the current ones, and they could be broadband based, so rather than relying on a telephone line to dial a number to the control room the control room will have visibility if you like of what's going on in that house all the time on line. So they're not having to wait for an adverse event to happen, for it to dial through to bring it to their attention, but they're able to look and see what's happening. And that could be using a camera and it could be using infra-red technology… all sorts, but it will change, I'm pretty convinced of that.

Essén (2008) rightly points out, however, that surveillance and control are integral parts of care and as such, they are both conceptually and empirically difficult to separate. In other words, whether care is inter-personal or technologically mediated, *all care involves some form of monitoring and observing;* disentangling care monitoring and surveillance is thus extremely difficult (Milligan 2009). Essén further argues that e-surveillance as in telecare home monitoring, need not be analysed as a panoptic ambition to control, since some users experience it as friendly and caring. Users may also see it as a way of avoiding more proximal forms of control as in residential care institutions for example. It is important to note, however, that Essén's study was based on a telecare trial in which an additional service was provided to twenty users who were then the subject of a close evaluation; under such conditions it is perhaps unsurprising that they experienced the trial as caring. The issue of how surveillance and monitoring should be viewed is, we would argue, contingent on both the user-context and the agency of the surveilled subject.

In her discussion of policy on old age, Pickard frames care practices as 'technologies of the self' (after Foucault) where such practices and techniques combine both 'repressive and productive roles' (2009, 68–69). Given that telecare technologies are aimed largely at older people deemed to be at risk in their own homes; that these systems collect and aggregate data which then afford new forms of subject and visibility; and given that they are set against the alternative of the residential institution; it is difficult (we maintain) not to see such systems as examples of governmentality where both involvement and resistance are at play. We do not then, view care technologies as being somehow either 'good' or 'bad', rather we argue the need to explore their potential whilst at the same time recognising their limitations. It is for this reason that we felt it important to engage with older people who had not yet come within this category, offering them the opportunity to reflect on those forms of NCT would be most acceptable to them and why.

Care Technology and Ontological Security

Home is an important site of ontological security (Saunders 1990; Depuis and Thorns 1998). It is the spatial context in which the day to day routines of everyday life are performed and where people feel free from the surveillance of the modem world. Older people within our study thus made clear distinctions between those care technologies they felt to be intrusive and unwelcome in the home, hence likely to reduce their personal agency, and those they felt supported their sense of safety and independence. In part this is wrapped up in their perceptions of the underlying purpose of NCTs; whether or not care was performed through these technologies; and consequently who, in their view, was most likely to benefit from their implementation. For some, emphasis on NCTs was less about the performance of care and more about managing risk and the increasing demands being placed on paid care and emergency services. As one citizens' panellist put it:

> Frank: ...I can see most of the benefits of this [telecare] being a benefit to the supplier of care or the service. I'm struggling to find the benefits for the cared for or the patient or the customer or whatever. ... they're managing the risk of having to turn out an ambulance to scrape you off the floor if you've fallen, so they're managing the risk.But caring for you to stop you falling in the first place, the system, telecare's not doing. No, risk management is not the same as care. They're aimed at two different people. Care is aimed at us, the patients. Risk management is aimed at the provider of the service.

Yet this is not the whole story. As with environmental control systems, the self-activated nature of wearable devices such as pendant alarms were viewed as not only contributing to an individual's sense of security and independence, but as increasing agency by locating the decision to activate the alarm firmly in with the hands of the older person him- or herself:

> Laura: I've had a pendant for must be seven or eight years, since my husband died. I didn't get one in the beginning, I was so much more capable really and thought, no I don't want a pendant and then eventually I thought, no, you're in the house on your own and at nights in the bedroom you can press it.

Many of the current generation of wearable devices available in the UK, however, are subject to limitations. Older people participating in our study for example, noted that while pendant alarms operated successfully within the immediate environs of the home or garden, their range was limited so did not extend much outside the home (for example communal staircases in multiple occupancy buildings). Despite their limitations, as the following citizens panel excerpt reveals, wearable, self-activated technologies were viewed by participants as providing a greater sense of independence and ontological security within the home than other forms of NCTs.

> Laura: well before I came out this morning I picked my pendant up off the bedside cabinet and took it downstairs with me, I wasn't going to wear it but as soon as I come in it's on the table the telephone table and ... take my things off, put the pendant on.

Brigit: I do that
Laura: And I feel safer with that.
Elsie: You see I don't think like... like the fall sensor, the pendant round your neck, I don't think they would make you more isolated, I think you'd just feel more secure and I think that those two
Susan: I think you're right, they give you more independence

Of course not all wearable devices are self-activated, other more technically sophisticated systems involve GPS tracking that enables monitoring to occur at significant distances from the older person's home (Tirado et al 2009). Call-centre workers within our study also pointed out that the sensitivity of wearable (automatic) falls devices was such that they could be triggered when an older person fell asleep and or merely slumped in a chair. Nevertheless, care professionals charged with implementing telecare systems suggest that the 'simple' reassurance of telecare is likely to promote greater independence than human-centred care which, some have suggested, may act to create dependence. As one social worker noted:

'People coming in can work on you psychologically – that sort of sense of dependence can be created. Whereas if you can call on a person [i.e. a call-centre operator] or this object – whatever it is we have in the house – [it] gives you the reassurance that contact will be available.'

Such a view points to growing debate about the nature of dependence, independence and interdependence – in particular, dependent on, or independent from, what or whom? (see for example, Thomas 2007; Schillmeier 2008; Milligan 2009). Indeed, we have argued elsewhere that that rather than reducing dependence, care technologies create new forms of complex socio-technical dependence, forms which are common to many social groups – some of which are not identified as 'vulnerable' (Mort et al. 2009).

NCTs, Home and the Changing Nature of Care Interactions

The discussion so far suggests that NCTs may be affecting the nature of care interactions with consequences for those involved in the implementation and performance of care. Importantly, this raises questions about whether the benefits of growing dependence on NCTs are likely to outweigh that of dependence on human caregivers – and whether this is a desirable outcome. This is exemplified in the following excerpts from a group discussion with older people:

Laura: you know, this is what we ought to be discussing, the difference between all these technological things and personal care, I think that is more important than all the gadgets that you can have in your house – it's the personal care, yes. and:

> Karen: oh, now how's a district nurse going to do a dressing then over the tele-line?
>
> Betty: no, you can't do things at long distance like that... the personal touch is very important, personal communication.
>
> Tessa: someone who cares, someone who cares about you, someone who really cares.
>
> Sally: what a terrible existence that would be, it's just a worry that technology would be there to see that you're alive, but you know... you're not....

As the above excerpt suggests, the giving and receiving of care requires human contact; technology itself cannot care, rather it offers a conduit through which new forms of human-centred care can be managed and delivered. In the context of shrinking care budgets, older people within this study have expressed concerns that these new forms of care could result in a decrease in social contact, increasing the potential for frail older people to become increasingly isolated within their own home. Asked what risks and benefits NCTs might pose for older people, one citizens' panel commented:

> Sarah: loneliness
> Murmurs of agreement
> Sarah: yes, no human...
> Karen: total isolation.
> (long pause)
> Kathleen: definitely
> Sarah: depression I should think
> I2: caused by...?
> Sarah: loneliness
> Kathleen: loneliness
> Sarah: and no-one to talk to, or a feeling of inadequacy
> Karen: and frustration as well if they're only going to get an electronic voice.

Research thus points to the need to take seriously older people's ongoing and ever-changing need for meaningful human interactions. So whilst NCTs are often cast as beneficial in facilitating older people's ability to remain at home, any reduction in personal care and concomitant increase in 'distance caring' made possible by telecare and web-cam technologies must recognise those aspects of care that remote systems are unable to address. Morris et al. (2003) for example, found that older people with varying states of cognitive decline felt very strongly about loneliness and the need to maintain social ties, arguing that meeting these social needs is central to older people's health status. Given that one of the fears most commonly expressed by older people, both in our own research and elsewhere, is that care technologies designed to monitor physical health status in the home might also result in reduced home visits from clinicians and carers, it is important to recognise how this might impact on older people's social and emotional well-being.

So on the one hand, NCTs can be seen as having an important role to play in enhancing the ability of older people to more effectively manage their lives within their own homes, whilst on the other they have the potential to further exclude and isolate. Understanding that NCTs should be considered as an *aid* and not a solution to growing demands for care becomes increasingly important when we also take into account the additional – and perhaps less overtly recognised – role that paid and unpaid carers play in the home. The cleanliness of the house, the need for repairs, the state of the garden and so forth, are all aspects of an older person's ability to deal with the general upkeep of their home that are often monitored by paid and unpaid care-givers entering the home, but which cannot be picked up by remote monitoring. As one older user of these technologies put it, 'what about "one picture's worth a thousand words?" If they just look at you they can tell very often, far better than talking to you [on the telephone]' (Roberts and Mort 2009, 153).

The Shifting Topology of Care

Many NCT systems rely on distant others to organise care. When a home sensor triggers an alarm, a call-centre operator contacts the user to find out whether the situation requires a response or whether it is a false alarm. Call-centres can be situated at some distance from the older person's home, where there may be little or no knowledge of the local environment. Yet locally-based call-centre operators in our own study highlighted the importance of this local knowledge, of the older people using NCTs and of their homes. In some instances call-centre operators had even met the older person in a related professional capacity; for example when visiting to install telecare equipment or acting as a warden in a sheltered housing or assisted living scheme.

This distancing of care clearly has a major impact on what care feels like to an older person. We have described elsewhere older participants' fears that telecare could be taken 'offshore' as has occurred with other teleservices, potentially causing serious communication problems (Roberts and Mort 2009, 146). We have also noted above concerns that the introduction of physical distance might act to destroy caring relationships altogether. Our observations of call-centre work and interviews with call-centre operators, however, reveals that operators perform a much broader caring role than the functional triage-based role outlined in telecare policy and technical literature (Audit Commission 2004; DoH 2005), or indeed that which is imagined by potential telecare users. Operators often have ongoing relationships with those using the system (although this is becoming increasingly limited as the numbers of individuals managed by any one call-centre grows) and employ a range of conversational techniques to provide a 'verbal care-substitute'. This is clearly illustrated through a conversation with two operators about their work:

> Anita: …same with the bed sensor, it tells you when they get out, and you do try and speak to the people just to make sure they haven't fallen … There's one

lady, she wanders all night long. Just moving herself. She has bad nights, she
doesn't sleep very well.
Bev: We should sing lullabies to her. I usually say 'It's too early now, get back
into bed, night night, God bless'. And they like you to talk to them like that,
[when]... you're lovely with them. You can talk to her.
Anita: You say 'We're having a bad nightmare, let's have a brew!' [chuckles]
I: Really?
Anita: Yes, she'll go and, because she's capable, make herself a cup of tea.

Whilst NCTs cannot provide physical care (indeed, their supporters and providers
strenuously argue that they should have no impact on the number of home visits
made to any particular client), the introduction of distance to a landscape of care
designed around ageing in place means adjusting to new forms of relationships,
including those with people an older person may only ever 'meet' on the telephone.
Telecare also produces new forms of caring work for operators who have to manage
their own feelings about being physically removed from those they are trying to
help. In the following quote, an operator describes her sense of helplessness in
being limited to this verbal form of care:

Bev: ...And you feel so helpless through the night if you get a call and suddenly
they'll say 'I want a wee.' Well you can't get the paramedics to go and get them
out to have a wee and they might not have any [other people to call upon]... I
mean we've got key safes to let paramedics and that in, but sometimes there is
nothing you can do, there isn't is there? There's nothing you can do, but try and
talk them into getting out of bed. I've done that before today.

This distancing of care through NCTs however, can also create difficulties for
older people. The hearing loss that often accompanies the ageing process can not
only create difficulties for some in responding to auditory-based technologies but
frustration at some call-centre care workers' lack of understanding of how to deal
with this issue. As one citizen's panellist commented:

Elsie: I find this, I mean I have two hearing aids and in fact I've just bought an
amplifier for me phone because I struggle on the telephone. You know, somebody
will ring up, somebody official, and you'll say ' I can't hear you, will you speak
up, I'm deaf' and they'll say 'oh right' and then immediately after they've said
right, their voice goes back down again.

For others, this goes beyond the problems arising from hearing impairments
to the difficulties and frustrations of dealing with NCTs designed around the
disembodied and dis-placed voice.

Edith: we had one of these machines that you put on your… if you've got a shelf in the hall, as you approach the front door it gave you a list of instructions. "Before you open the door you do this, you do that and the other…"
laughter
Edith: and after one week we threw it in the bin because it drove you up the wall because it worked every time you went to the front door whether you went out or came in or something and it literally drove us up the wall, so that was a total waste of money.
Susan: it's easier to stick a post-it note on the door (laughs).

Hence, the impact of NCTs cannot be seen as minor shift in the landscape of care. We maintain that their effect on the distancing of care, particularly at times of crisis, but also on a more mundane level, seriously alters the topology of care-giving and receiving.

Discussion: Care and Home – Dis-placing Place?

The home represents a personalised, private territory where habitual 'modes of operating' are invented, organised and performed in ways that gain defining value. As a site of embodied memory, identity, self-expression and socialisation it is both materially and emotionally dense. Any understanding of the relationship between NCTs, home and ageing in place thus needs to recognise that home is not simply a physical or spatial setting, but is also an environment of experience and meaning in which transactions between people, place and things are constructed and reconstructed over time and circumstance. Such a setting is not only seen as conducive to an individual's personal well-being but represents a site within which that individual can maximise his or her personal agency.

For older people requiring care and support to manage their daily lives, the home also exemplifies 'a protected place' where the sick or ageing body can find refuge and care away from the pressures of the social [collective] body. That is, the home is experienced as a site of ontological security. The performance of care through self-activated NCTs such as pendant alarms is clearly seen by older people within our study as holding the potential to enhance their sense of home, ontological security and personal agency. But NCTs are also contributing to a new topology of care – one that distances care and brings new actors into the care network. Call-centre operators, care providers, care professionals and family carers are all drawn into new care relationships within the home and across both virtual and physical space that older people experience in both positive and negative ways. This new topology of care is having a major impact on what care feels like to older people.

Further, for many older people in western societies, personal agency is directly linked to having one's own 'front door' – a material symbol that demarcates the boundary between public and private with all the rights inferred from owning or renting that dwelling (Percival 2001). The front door is emblematic of the older person's ability not only to decide who to exclude from the home and who to allow entry, but also the freedom to make decisions about how life is lived within that home. Yet as this chapter illustrates, as older people become increasingly frail, the home comes to represent a 'theatre of operation' in which a multiplicity of care functions and practices take place. The effects of new care technologies on policies and practices designed to support ageing in place thus highlight the spatial dialectic of home – one that not only reinforces a growing porosity between public and private, workplace and home, but which also points to shifts in the performance and power-dynamics of care that are occurring both within and across the home (Milligan 2003).

We would also argue that such technologies represent emerging cracks in the door that demarcates the boundary between public and private space – and along with it, older people's choices about who to exclude and who to allow entry into the home. Indeed, in this chapter we have illustrated how NCTs act to circumvent 'the door', opening up new ways of entering the home through sensors, webcam and telecare technologies. Such technologies enable care professionals and call-centre operators to 'appear' in the home albeit without a physical presence. To date such appearances still require the older person's decision to actively participate or respond, however the vision of a new generation of NCTs described by one manufacturer in this chapter would enable care-call operators to 'appear' at will. The emergence of GPS-based systems and internet-based technologies also enable care-givers to monitor and map the older person's movements within and beyond the home from remote 'care' settings (Milligan 2009; Tirado et al. 2009). Whilst these kinds of development are couched in terms of prevention and risk minimalisation, inevitably they come at the cost of a loss of the older person's personal agency and power to exclude. Hence NCTs may also act to dissolve interior and exterior distinctions – as one older participant put it, 'technology has no bounds has it?'

In sum then, despite the relative physical unobtrusiveness of remote monitoring and surveillance technologies in the home, particularly in comparison to adaptive devices such as hoists, lifts, rails etc. NCTs may have significant effects on the meaning and 'sense of home' and how older people identify with it. The risk is that over-reliance on some forms of NCTs or the development of more intrusive NCTs may act to reshape it – stripping the home of its personal meaning and identity. Ironically, such developments might be seen as having more in common with the institutionalisation of the home than the facilitation of any preconscious sense of setting that can enhance ageing in place.

References

Angus, J., Kontos, P. C., Dyck, I., McKeever, P. and Poland, B. 2005. The personal significance of home: habitus and the experience of receiving long-term home care. *Sociology of Health and Illness*, 27, 161–187.

Augé, M. 1995. non-place: introduction to an anthropology of supermodernity, (translated by Howe, J.). London: Verso.

Audit Commission 2004. *Implementing Telecare. Strategic analysis and guidelines for policy makers, commissioners and providers*, Public Sector National Report; www.audit-commission.gov.uk/olderpeople [Accessed November 2009].

Bayer, S., Barlow, J., and Curry, R. 2007. Assessing the impact of a care innovation: telecare. *Systems Dynamic Review*, 23(1), 61–80.

Blythe, M. A., Monk, A. F. and Doughty, K. 2005. Socially dependable design: The challenge of ageing populations for HCI. *Interacting with Computers*, 17, 689.

Blunt, A., and Dowling, R. 2006. *home.* London: Routledge.

Cowan, D. and Turner-Smith, A. 1999. The Role of Assistive Technology in Alternative Models of Care for Older People (Appendix 4) in *With Respect to Old Age: Long term care- Rights and responsibilities: Research Volume 2: Alternative models of care for Older People, Royal Commission on Long Term Care* edited by S. Sunderland. London: HMSO.

Davidson, T. 2009. The Role of Domestic Architecture in the Structuring of Memory. *Space and Culture*, 12(3), 332–342.

De Certeau, M., Giard, L., and Mayol, P. 1998. *The Practice of Everyday Life Volume 2: Living and Cooking.* Minneapolis: University of Minnesota Press.

Demeris, G., Rantz, M. J., Aud, M. A., Marek, K. D., Tyrer, H. W., Skubic, M. and Hussam, A. A. 2004. Older adults' attitudes towards and perceptions of 'smart home' technologies: a pilot study. *Medical Informatics and the Internet in Medicine*, 29, 87–94.

Department of Health 2008. *The Preventative Technology Grant.* Available at: www.dh.gov.uk/en/Publicationsandstatistics/Publications/PublicationsPolicy AndGuidance/Browsable/DH_5464107 [accessed May 2009].

Dupuis, A. and Thorns, D. 1998. Home, home ownership and the search for ontological security. *Sociological Review*, 46(1), 24–47.

Essén, A. 2008. The two facets of electronic care surveillance: An exploration of the views of older people who live with monitoring devices. *Social Science and Medicine*, 67, 128–136.

Giddens, A. 1990. *The Consequences of Modernity.* Stanford CA: Stanford University Press.

Lansley, P., McCreadie, C., Tinker, A., Flanagan, S., Goodacre, K. and Turner-Smith, A. 2004. Adapting the homes of older people: a case study of costs and savings. *Building Research and Information*, 32(6), 468–483.

Lianos, M. 2003. Social control after Foucault. *Surveillance & Society*, 1(3), 412–430.

López, D. and Sánchez-Criado, T. 2009. Dwelling the Telecare Home: Place, Location and Habitability. *Space and Culture*, 12(3), 343–358.

Magnusson, L. and Hanson, E. J. 2003. Ethical issues arising from a research, technology and development project to support frail older people and their family carers at home. *Health and Social Care in the Community*, 431–439.

Miller, E. A. 2001. Telemedicine and doctor–patient communication: an analytical survey of the literature. *Journal of Telemedicine and Telecare*, 7(1), 1–17.

Milligan, C. 2000. Bearing the burden: towards a restructured geography of caring. *Area*, 32, 49–58.

Milligan, C. 2003. Location or Dis-Location: from community to long term care – the caring experience. *Journal of Social and Cultural Geography*, 4, 455–470.

Milligan, C. 2006. Caring for older people in the 21st Century: notes from a small island. *Health and Place*, 12, 320–331.

Milligan, C. 2009. There's No Place like Home: People, Place and Care in an Ageing Society. Aldershot: Ashgate Geographies of Health Book Series.

Morris, M. E. 2005. Social Networks as Health Feedback Displays. *IEEE Internet Computing*.

Morris, M., Lundell, J., Dishman, E. and Needham, B. 2003. New Perspectives on Ubiquitous Computing from Ethnographic Study of Elders with Cognitive Decline. *UBICOMP*, LNCS 2864, 227–242.

Mort, M., Roberts, C. and Milligan, C. 2009. Ageing, Technology and the Home: a critical project. *Alter: European Journal of Disability Research*, 3(2), 85–89.

Percival, J. 2001. Self-esteem and Social Motivation in Age-segregated Settings. *Housing Studies*, 16(6), 827–840.

Pickard, S. 2009. Governing Old Age: The 'Case Managed' Older Person'. *Sociology*, 43(1), 67–84.

Roberts, C. and Mort, M. 2009. Reshaping what counts as care: older people, work and new technologies. *ALTER: European Journal of Disability Research*, 3, 2138–2158.

Rowles, G. 1978. Prisoners of Space? Exploring the Geographical Experiences of Older People. Boulder, Colorado Westview Press.

Rowles, G. D. 1993. Evolving images of place in aging and Aging in Place. *Generations*, 17(2), 65–51.

Rubinstein, R. 1989. The home environments of older people: a description of psychosocial processes linking person to place. *Journal of Gerontology*, 44, S45–53.

Sacco, V.F. and Nakhaie, M.R. 2001. Coping with crime: An examination of elderly and non-elderly adaptations. *International Journal of Law and Psychiatry*, 24 (2–3), 305–323.

Saunders, P. 1990. *A Nation ode Home Owners*. London: Unwin Hyman.

Schillmeier, M. 2008. Time-Spaces of In/dependence and Dis/ability. *Time & Society*, 17, 215.

Schillmeier, M. and Domènech, M. 2009. Caring and the Art of Dwelling: Bodies, Technologies and Home. *Space and Culture*, 12(3), 288–291.

Tinker, A., McCreadie, C., Stuchbury, R., Turner-Smith, A., Cowan, D., Bialokoz, A., Lansley, P., Bright, K., Flanagan, S., Goodacre, K. and Goodacre, P. H. 2004. *At Home with AT: introducing assistive technology into the existing homes of older people: feasibility, acceptability, costs and outcomes.* London: Institute of Gerontology King's College London and the University of Reading.

Tirado, F., Callén, B. and Cassián, N. 2009. The Question of Movement in Dwelling: Three Displacements in the Care of Dementia. *Space and Culture*, 12(3), 371–382.

Thomas, C. 2007. Sociologies of Disability and Illness: Contested ideas in disability studies and medical sociology. Hampshire: Palgrave Macmillan.

Tracy, C. S., Drummond, N., Ferris, L. E., Globerman, J., Hebert, P. C., Pringle, D. M. and Cohen, C. A. 2004. To tell or not to tell? Professional and lay perspectives on the disclosure of personal health information in community-based dementia care. *Canadian Journal on Ageing-Revue Canadienne Du Vieillissement*, 23, 203–215.

Tuan, Y.-F. 2004. *Home, in patterned ground: the entanglements of nature and culture*, edited by S. Harrison, S. Pile and N. Thrift. London: Reaktion Books, 164–65.

Tucker, S., Hughes, J., Burns, A., and Challis, D. 2008. The balance of care: Reconfiguring services for older people with mental health problems. *Aging & Mental Health*, 12(1), 81–91.

Wiles, J. 2003. Daily geographies of care-givers. *Social Science and Medicine*, 57, 1307–1325.

Wiles, J. 2005. Home as a new site of care provision and consumption, in *Ageing and Place: Perspectives, policy and practice*, edited by G. Andrews and D. Phillips. London: Routledge, 79–97.

Wiles, J., Allen, R., Palmer, A., Hayman, K., Keeling, S. and Kerse, N. 2009. Older people and their social spaces: A study of well-being and attachment to place in Aotearoa New Zealand. *Social Science & Medicine*, 68, 664–671.

Chapter 2

The Securitization of Care Spaces: Lessons from Telecare

Daniel López

Introduction: New Technologies of Community Care for an Aged Society

Telecare is usually presented as a new technological solution to care-giving. Information and Communication Technologies (ICT) and other monitoring technologies are seen as offering the possibility of assessing the needs of people with a high degree of dependency, and of giving them medical or social assistance in their home as fast as possible.[1]

But, telecare is turning out to be not just a technological innovation but also a new means to reconfigure the organization of care delivering with important economic, social and political consequences. According to telecare developers and policy-makers, telecare may become a solution to cope with the increasing and long-term care necessities arising from transformations in the family, the breakdown of the traditional care institutions, the aged population, and the prevalence of chronic diseases. Thus, telecare seems to put into practice a new kind of care policy, based on community care and ageing-in-place, as a strategy to foster the independence and wellbeing of both the cared and the caregivers while cutting the costs of care delivering.

The aim of this chapter is to offer a critical diagnosis of the community care that telecare technologies are bringing about by establishing a dialogue among the following three areas: an ethnography conducted between 2003 and 2004 in a telecare service located in Catalonia/Spain (see López 2008; López and Domènech 2008; López and Domènech 2009; López and Sánchez-Criado 2009), the Foucauldian literature on governmentality (Dean 1999; Gordon 1991; Rose 1999) and the so-called Surveillance Studies (Lyon 2006). Our intention is to shed light on the differences and continuities between telecare and traditional caregiving institutions. By drawing on Foucault's genealogy of power we will show how an analysis of this remote care system results in a definition of community care as a shift from a disciplinary regime to a regime based on security. This

---

1 Although tele-alarms with personal triggers, is the most widespread telecare systems in most of the European Countries, telecare's morphology and functions are changing rapidly as technologies such us telemedicine devices, sensors and simulation technologies are being incorporated to it.

approach contrasts with the understanding of remote care as being a mere shift in the place of care (such as from the institution or the home to an open, reticular space of care), or as a care practice that patronizes elderly people and fosters their dependency to another form of care whose main goal is to improve their autonomy and wellbeing. The political implications of this shift will be discussed

Telecare as a Care-delivering Solution

Telecare is often presented as a service that can cut long-term care costs by transforming the traditional organization of care delivering. In order to cope with the increase in long-term care needs, care giving responsibility has been distributed among different entities that ideally should work in unison: families, neighbors, associations, NGOs, public administration, insurances, technology developers and service providers. It is a communitarian care-giving logic (Mira Johri 2003; Winthereik and Vikkelsø 2005) characterized by three main features:

a. care-giving is not the duty of a single entity anymore but the duty of an array of diverse actors that must work in coordination with each other, not according to a shared vocation but towards the accomplishment of efficiency and quality results;
b. the involved actors are not proper institutions anymore, but de-bureaucratized and flexible and permeable organizations;
c. the State is no longer considered as an all-powerful subject in policy-making, instead networks of public and private agents are made responsible for policy-making and for meeting certain measurable objectives (Dubet 2002).

Regarding this, telecare and more generally e-health appear as an intensification of this process of *communitarization* of care (Bashshur et al. 2000). Actually, the main goal of telecare services is to become the entrance point into the community's network of aid resources.

Moreover, telecare also seems to put into action most of the principles of 'the ageing-in-place' paradigm. This paradigm points out the constant adjustment between the ageing processes and the places where these processes are located as being essential for good ageing.[2] Moreover, it consequently sets up a political

---

2 As it is expressed in the following quote, ageing-in-place is a therapeutic achievement to increase the quality of life of the elderly people as well as a current tendency in policy-making on aging and care delivering: "To become more responsive to the needs of older adults for maintaining a self-identity of competence and feeling "in place", aging policy and practice must incorporate three main themes: (1) emphasis on consumer direction and consumer values; (2) preference of residential, home-like, "caring" settings over institutional, hospital-like, "curing" settings; and (3) establishment of a continuous LTC system both able and willing to provide supportive services that facilitate aging in place" (Leith 2006: 331).

goal for community care. According to the ageing-in-place paradigm, ageing in the usual context (mainly in the home) helps to maintain autonomy for as long as possible, i.e. to allow for the capacity to control one's own environment.

In this vein, telecare is seen as a technological catalyst for implementing a community and aging-in-place care system that can cope with the increasing long-term care needs of our societies.

The Telecare Panopticon

I would like to introduce the discussion concerning the way this community-based and ageing-in-place care delivering model is enacted and shaped by telecare by presenting an excerpt of the fieldwork regarding the process of attending to a call in the call centre.

> There is a new call in the telephone inbox at the alarm headquarters. According to the information that appears on M's computer screen, it is a call that user 432567 has placed from the button. M selects the call to answer it, and as she gets a line to talk the information on user 432567 is shown on the screen: her name is Mrs J, she lives alone in Barcelona and is 73 years old. M asks loudly and clearly: "Good day, Mrs J. How are you today? How do you feel?" When she hears no answer, she opens up the record of calls to see how the previous calls from Mrs J have been codified. Several calls are shown with a code made up of an A or S and two digits showing whether the call was incoming or outgoing, and another three-digit code – the kind of action that was performed. Most of the calls are outgoing courtesy calls and errors (A33–100 or A37–100). This means that there has been no recent emergency. Still, when M more thoroughly explores Mrs J's file, she begins to get nervous: Mrs J is obese. She could have fallen far from the terminal and perhaps cannot get up. M quickly chooses the first main contact that appears in the database and calls it. Mrs J's son answers and she tells him the situation. To avoid alarming him she says that his mother's alarm has gone off, but that it might be because she is hanging out the laundry and has pressed the button without meaning to. But right then, another call appears in the inbox with user code 432567, and even before choosing it M changes her tone of voice and tells Mrs J's son to go to his mother's house quickly because she might have hurt herself. M then codifies the call she made to Mrs J's son in the database. Meanwhile, she chooses the new incoming call, takes off the earphones and very loudly asks if Mrs J has fallen or needs help. In the background she can hear Mr J answer. At that point, she leaves the line open in order to monitor what is happening at the home, while she once again calls the main contact and tells him that she has fallen. M once again codifies the call in the database and hangs up the line. She says to herself: "Come on, you're not done yet". Five minutes later, she can hear the son entering Mrs J's house. He tells M that his mother has fallen, she has hurt herself and is bleeding. M immediately opens up a window

in the database and calls 061 for emergencies. There she is asked for Mrs J's information and M tells them that she is an overweight user who has fallen, she is bleeding and that her son is with her. The ambulance service says that they will call the home and that M will have to hang up to free the line. M codifies the call to 061, hangs up and then tells Mrs J's son to hang up because a doctor will be calling him to give him instructions as to what to do, and that an ambulance is on its way. She asks that he then push the button again so that she can continue to help them. However, the son seems to be a bit disoriented and unsure what to do in this situation. M tells him that the most important thing is not to touch anything and that at most he should cover the wound with a clean gauze cloth so that it stops bleeding. She also suggests that he try to talk to his mother to prevent her from fainting, and not to move her. She codifies the call and hangs up. In the meantime, M has been noting down what kind of calls she has placed and what has been happening in strict chronological order on one of the call follow-up reports. When she has finished, when she sees that neither the son nor the EMTs from the ambulance have called, she chooses the home phone number in Mrs J's file and calls. The son answers and tells her that the ambulance has just arrived and is taking his mother to the hospital. M offers to tell his brother (the other main contact that appears in the database) and he agrees. M codifies the call and chooses the second main contact in order to call him. His wife answers, but M tells her about what has happened, stressing the fact that everything is fine. She then codifies the call, finishes noting down what she has done on the call follow-up form and then forwards the report to another operator.

This example of care at a distance usually triggers questions concerning: a) its materiality – are we witnessing a new form of care delivering mainly based in information management rather than hands-on-care? b) its spatiality/temporality – are we witnessing the rise of a new care space characterized as a network where the caregivers become immediate aid resources managers? c) concerning the object of care –are we witnessing a care practice focused on monitorizable behaviors?[3] Indeed, these questions express three conceptual polarities through which the impact of telecare and other e-health technologies in community care are usually approached: body vs. information, places vs. networks, moral subjects vs. monitorized behaviors (May et al. 2005; Mort et al. 2008; Nettleton 2004; Percival and Hanson 2006). But these polarities have not only been used in analytical terms but also as a means to explicitly express an ethical concern over telecare and other e-health technologies. The informatization of care delivering, the organization of care delivering in delocalized and ubiquitous networks and the preeminence of monitoring as a much more important care activity throughout different caring settings leads to questions about the definitions of ourselves in terms of being in need of care and taking

---

3   An illustrative example of this is the British Telecom Care in the Community Project carried out in Dundee and Liverpool to model the lives of older people in order to understand, anticipate and respond to their home based care needs (see Hine et al. 2005).

care of others, and thus about responsibility with respect to others and about our freedom and privacy. Even though these ethical questions clearly connect care and power, the contribution of surveillance studies[4] or governmental studies to a critical analysis of telecare developments has been paradoxically minor. Precisely, this is something we want to address in this chapter.[5]

Drawing on surveillance and governmental studies, the three before mentioned polarities concerning telecare might actually be expressing a structural shift from a society made of panoptic institutions to one made of post-panoptic control arrangements (Haggerty 2006).

As we are going to explain in depth, this transition from the former to the latter appears to be inscribed in telecare's operations and draws on a particular understanding of community care. In panoptic institutions, the interiorization and autonomization of the disciplinary order is the result of a very physical surveillance; whereas in telecare's case, surveillance seems to not mainly be addressed to the body as the preferred surface of power. Even though in telecare's case the body might continue to be the target of a surveillance that springs from itself much more intensively due to new digital technologies, the panoptical sight as main surveillance procedure has been replaced by practices of codification, decodification and simulation of information. In our case, the telecare operator evaluates what is happening to the user through different information concerning drugs, diseases, relatives, contexts and other past events registered in a database. It results in the user not feeling a coercive surveillance. Accordingly, our analysis of telecare might coincide with those theories that propose that vision mechanisms of surveillance are being replaced by mechanisms of forecasting and precaution based on a functional link between acceleration of information transmission and an increase in the capacity for supervision. Contemporary surveillance assemblages are formed by different means of gathering, sorting, recalling and processing information (including information blocking or jamming technologies) (Bogard 2006).

According to Haggerty and Ericson (2001), while disciplinary institutions depended on the totalization and homogenization of control systems, in current contexts "surveillance is more decentralized, less subject to spatial and temporal constraints (location, time of day, etc) and less organized than ever before by the dualisms of observer and observed, subject and object, individual and mass."

---

4   This field is characterized by studies that usually face these polarities concerning the materiality of power, the spaces of power and the process of constituting subjectivity entailed when comparing new technologies such as internet (Lyon 1994), CCTVs (Graham and Wood 2003) and screening technologies (Graham 1998) with the analysis of modern institutions made by Michel Foucault (1977).

5   Actually, even though this is not our main aim, this critical reflection on telecare might be considered a contribution in the Surveillance Studies because it is focused on technologies considered as 'new' and because almost none of these studies are about assistive technologies such us telecare or telemedicine (some exceptions are Essén 2008, Percival and Hanson 2006).

(Haggerty and Ericson 2001: 102) This description might be applicable to the case of telecare. Surveillance in telecare does not totally depend on a direct sight of the users, but rather on the control and coordination of different aid resources that have to keep watching each other through different assessing and auditing procedures. Thus, in the case of telecare, the hierarchical surveillance would be replaced by lateral control, and we would face a network rather than a pyramid. Indeed, it is precisely this feature which makes telecare and e-health different from traditional care delivering institutions (Bashshur et al. 2000). Care delivering currently involves connecting and coordinating emergency services, hospitals, GPs, insurance companies, relatives, and neighbors.

Regarding the process of constituting subjectivity, telecare might also fit within a post-panoptical regime because the constant adjustment of habits and spatialities in permanent transformation is much more important than the production of a specific, coherent and stable identity for the users. Lianos (2003) gives us a clue to understanding this shift through an analysis of the power relationships that alarms are enabling in public spaces.

> As a result, we must stop looking at the purpose and the consequences of institutional control as a prolongation of subjection, either in the sense of submission or in the Foucauldian sense of the constitution of the socialised human being as a subject. The broad, moral and disciplinary, modernist vision of control is no longer operational in the overwhelming majority of contemporary environments where collective interaction takes place. On the contrary, institutional control is about the 'de-subjectification' of the individual, who is being largely transformed into a fragmented user, since the object of control is to regulate exclusively the specific institutional shell of activity concerned each time. (Lianos 2003: 423).

If we follow Lianos (2003), surveillance in telecare would not aim to produce the normalized subjects (soldiers, students, parents or workers) of the discipline institutions, but to control and monitor specific habits deeply related with the function of telecare, for instance: monitoring if the user is wearing the pendant most of the time, pushing the button when it is needed or merely speak to the terminal from the proper distance, etc. These kinds of habits configure the space of the telecare and therefore become objects of a subtle and punctilious control.

Highlighting these changes concerning the materiality and spatiality of care and the subjectivation processes entailed helps us to understand how telecare developments are configuring community care. Nevertheless, I think that it is necessary to adjust and rethink some of the axioms that these studies usually use to conceptualize these changes. I would like to discuss two of these axioms before offering a critical understanding of community care drawn from the case on telecare.

Firstly, pointing out the post-panoptic features of telecare (information, networks and monitoring of behaviors) might entail hiding or ignoring characteristics present within telecare that are closely linked to traditional care settings. For instance, as we have seen in our research on telecare, corporal discipline is very necessary for the service. Telecare's incorporation into the home requires, on the one hand, assigning a specific time and place to behaviors, objects and events, and on the other hand, embodying a very physical and concrete disposition to act: wearing the personal trigger (normally a pendant) all the time while at home, only taking it off when you are leaving the home, pressing it whenever you need help and also once a month just to check that it is working correctly. Thus, what is really being installed in the home is a specific way of dwelling in which the personal trigger must turn into the *obligatory passage point* of any aid request, and, therefore, in which the homes must become a hearable and transparent spatial extension for the operators of the service (López and Sánchez-Criado 2009). Secondly, it is common to confuse the panopticon with the idea of a total institution.[6] The changes on materiality and spatiality of care, as well as on the process of constituting subjectivity involved, refer to the transition from the discipline-blockade to the discipline-mechanism but not to the disappearance of the panopticon. The importance of information management, the network organization of care and the monitoring of user's behaviors and habits at a distance are characteristic of disciplinary mechanisms inside or beyond the borders of total institutions, and this generalization of disciplinary mechanism is precisely what Foucault (1977) called panopticism.

Thus a panoptic society is not a society configured by confined spaces, but a society in which the normalizing surveillance is socially spread out through different micro-technologies. *Discipline and Punish* (Foucault 1977) is a book about the process of diffusion of discipline and not about enclosing it. Foucault firstly defined disciplines through the spatio-temporal regime of enclosing institutions such as prisons, workshops or schools, but he then proceeded to show how discipline moved into other types of spaces, connecting different institutions and setting up a broader and not necessarily enclosing space: "the social" (see Donzelot 1979, Rose 1996b). Thus, as he says, "Discipline sometimes requires enclosure, the specification of a place heterogeneous to all others and closed in upon itself. It is the protected place of disciplinary monotony (…). But the principle of 'enclosure' is neither constant, nor indispensable, nor sufficient in disciplinary machinery. This machinery works space in a much more flexible and detailed way" (Foucault 1977: 141).

Actually, Foucault (1977) is describing a movement towards the externalization of disciplinary mechanisms that has three stages: leprosarium, which excludes in order to keep a community pure; quarantine against plagues, which is already a disciplinary mechanism but is exceptional in that it strives to fix any movement and to set up spaces of total surveillance in order to individualize the plague; and finally the panopticism, where the disciplinary mechanisms operate in different contexts

---

6 See for example Bogard (2006), "One way to describe the evolution from panoptic to post-panoptic systems is from territorial to deterritorialized forms of social control, for example, from guarded or confined spaces to digital networks" (Bogard 2006: 97).

and as invisible routines rather than as an exceptional solution to exceptional situations. Regarding this, Foucault insists that the panopticon would not have become a generalized technology of power (a technology capable of defining the whole society) if technologies of inscription like psychological tests, much more flexible and mobile, had not been involved (Rose 1996a). These flexible and mobiles technologies composed a surveillance assemblage (Haggerty and Ericson 2001) through which the old confined institutions where articulated and through which the panoptic surveillance went further them reaching the whole society.

However, the presence or absence of enclosing are not the most interesting features of the panopticism but the rationale of power that it embodies. Power is not exerted only over prisoners, patients or students, but over the relationship between people and things. The aim of the panopticon is to manage the network of people and things and its development (Miller and Miller 1987), and, in order to accomplish this purpose, technologies of representation such as reports and tests were totally necessary. They enabled the enforcement of a displaced, remote and simplified control over any multiplicity of elements. As Cooper (1992) has said:

> The whole idea of the Panoptic principle was to connect abbreviated representations – models, signs, summaries, and the like – with a many-layered imbrication of social, political, architectural and other factors in a kind of semio-technical hierarchy where the simplest term could represent the most complex series, where the most intricate details of institutional behaviour could be orchestrated to respond to the briefest command, the most peremptory signal! Today, this programme is being dramatically advanced by means of information technology. (Cooper 1992: 265–266)

The panopticon gives us a lesson on government. To govern properly it is vitally necessary to get a precise and continuous knowledge from these networks of things and people in order to, firstly, conduct these networks towards the purposes that have to be accomplished; and, secondly, to adapt these purposes to the state of these networks.

Telecare, thus, might be understood as a kind of panopticon that manages the life of the users at home. This panopticon works by producing different sorts of inscriptions: disciplined bodies such as user who know when to press the alarm and how to do it and mobile inscriptions such as files with details about the user that enable action to be taken at a distance from a more or less stable centre. The aim of the service is thus to configure a delivering care programme[7] by arranging and managing the highest amount of elements that are or might be affecting care delivering: from the spatial configuration of the home, the location of the nearest hospital, the relatives of the user and the municipal police.

---

7   This idea has been inspired by the notion 'programme of action' as used by Akrich and Latour (1992).

However, as happens with the panopticon, the networks of people and things are not stable but living entities. It is also necessary to constantly readjust the delivering care programme of the telecare service to the changes that these networks of people might undergo. It is necessary to check the battery of the device, if the personal trigger is working properly, check how the user feels and how their relatives or neighbors are (i.e. the primary contacts to be mobilized in case of emergency) and also check if the user's list of drugs has changed by calling once a month or by enquiring when the user presses the button in order to chat or by accident. As we have seen, even during an emergency call, all actions and events must be written down in the follow-up report.

In this vein, the delivery of care reveals a dynamic normalized process: a) a specific care delivering program is inscribed in different sorts of inscriptions –for example, there is a normative way to deal with any call embodied in specific protocols; b) this care delivering program is updated when put into practice – the way the operator is actually dealing with the situation that materializes itself in notes always brings about new differences that are registered along with recordings of the call and the call follow-up reports; c) a re-inscription of the care delivering program is produced by assessing if the inscription of program matches the inscriptions of the actual program – whenever there is a periodical internal assessment or a complaint from a user, the managers compare the protocols with the notes, call follow-up reports and recordings and then change the protocols or other implicated procedures. Summing up, telecare is a kind of inscription machine, a machine that translates any event or element of the user's life at home to much more stable codes and elements whose objective is to gather knowledge and to act rapidly and reliably (López 2008).

In precisely this respect, the compulsion to produce inscriptions and to re-inscribe any differences is the most salient characteristic of the panoptic administration and the signal that marks the beginning of a different understanding of telecare and community care. This impulsive control of any *difference*, which looks for a constant quality improvement of the service, reveals a new logic or rationale that goes further than the discussion about the panopticon/postpanopticon regime and enables us to construct a critical interpretation of the care-delivering model that telecare developments are bringing about.

From Discipline to Security

The constant work of inscription and re-inscription carried out in the Home Telecare Service has a specific goal: to improve the quality of the service and therefore, the users' satisfaction. But how is this quality established? The threshold is set up according to a risk calculus. The main goal of this constant work of inscription, actualization and re-inscription is to reduce risks and discover new risks.

The call codes and the intervention codes, as well as the protocols that are used in the call centre by the teleoperators, have been designed based on a principle of precaution. Every response from the service has been planned according to abstract risks. Whether the user is screaming out because they are lying down and is not able to get up or because the user is still in bed and apparently just complaining, the teleoperators must follow the protocol because it has been designed to minimize the risks of average medical emergency.

The notion of risk is very important because it changes the very essence of what an event is. As Ewald (1991) has said, "Nothing is a risk in itself; there is no risk in reality. But on the other hand, anything can be a risk; it all depends on how one analyses the danger, considers the event" (Ewald 1991: 199). Events are not necessarily identified by risk as dangers or threats, but are instead defined as a possibility or chance of loss or damage. Essentially, the 'risk operation' allows these elements of reality to be assigned new values. They stop being obstacles and turn into being possibilities (of gain or loss).

> To calculate a risk is to master time, to discipline the future. To conduct one's life in the manner of an enterprise indeed begins in the eighteenth century to be a definition of a morality whose cardinal virtue is providence. To provide for the future does not just mean not living from day to day and arming oneself against ill fortune, but also mathematizing one's commitments. Above all, it means no longer resigning oneself to the decrees of providence and the blows of fate, but instead transforming one's relationships with nature, the world and God so that, even in misfortune, one retains responsibility for one's affairs by possessing the means to repair its effects. (Ewald 1991: 206)

In our example, as the risk becomes independent from the specific threatening situation, the delivering of care changes. Rather than persons, situations and collectives, risk profiles are the object of care. Such a shift becomes possible as soon as the notion of risk is made independent from that of danger. A risk does not arise from the presence of a particular precise danger embodied in a concrete individual or group or in a specific situation. It is the effect of a combination of abstract factors that render more or less probable the occurrence of undesirable modes of behavior (Castel 1991: 287). This means that delivering attention is being ruled by specific risks (through protocols that establish what to do) and mechanisms that were developed to integrate, concurrently, unforeseen events or phenomena and thus permanently adjust how care is delivered.

But risk is also present in technology design, specifically in the adjustment between user and technology. The evolution of the personal trigger's design ended up being a pendant specifically designed against the risk of falling. As a result of different studies, unattended falls turned out to be the problem that this kind of alarm has to deal with. These types of fall are an increasing problem for the elderly as a result of normal physical deterioration and the fact that many older people live alone at home with a risk of falling and without receiving aid for long time (handicapped persons or those with chronic illnesses) (see also Thygesen and Moser, in this book). This population needs to ask for help to face many different types of situation. With this purpose in mind, the personal trigger was designed to be carried all the time without effort. One of the designs is a strap, like a wristwatch. However, according to the managers of the service, it has two problems: it gets dirty easily and some studies have revealed that if the person falls a certain way it is possible that they cannot reach the button. As a consequence, an alternative design was put forward taking into account that after a fall the button would more likely be pressed if the push button is a pendant rather than a strap. It was also considered a better design because the entailed risks, such as the pressing by accident, were not prohibitive. Even though it is very improbable, the risk of strangulation was considered the only real threat. To safeguard against this, the pendant was designed with a special lock that is released when it is under a certain amount of pressure.

The design and uses of the pendant were configured not so much through a normative conception of what telecare and a user should be, as through what might happen, i.e., the risks entailed. This is why the pendant's design is open. There are always questions arising about how the technical design of the devices might or might not cover new emergent risks. For example, what would the risks be if the users decided to customize their pendant by turning it into a brooch clipped to a lapel? Are these risks being taken into account in the technical design of the pendant?

The importance of risk in either the care delivering process or the design of telecare technologies puts security at the foreground of telecare. Security appears as a commodity as well as a procedure. Elderly people usually get a telecare service in order to count on somebody that will always know how to help them whenever something dangerous happens at home. Telecare is there just in case. Telecare is dealing with a risk, mainly the possibility of falling at home completely alone. But in order to provide safety to elderly people and their relatives, the service must work reliably. The users, firemen, ambulances and other aid resources are connected and work together remotely due to the production and arrangement of different kinds of inscriptions, but this cooperation is also achieved because more inscriptions are produced to monitor and 'readjust' any task or procedure. The service should be sensitive to the risks entailed in its functioning.

This risk management is what Foucault (2007) defines as security and enables a social diagram to be drawn that connects telecare with other contemporary systems. In *Security, Territory, Population*, Michel Foucault (2007) defines security as operating through practices that were also present in disciplinary mechanisms (such as training) or in the law (such as punishment and incarceration) but which are being redefined by a different logic. Law proceeds negatively: what is not forbidden is allowed. Its correlative is the punishment when the law has been broken. On the other hand, discipline codifies indistinctively what is allowed and what is banned, not from the standpoint of the latter, but from the angle of what should be done. Discipline is a positive and productive power, thus, it does not work through death, by the right to kill or hurt in the case of a crime, but instead moulds life according to certain patterns. What must be done is perfectly regulated.

Security, instead, does not operate from the point of view of what is forbidden or what should be done, but from the point of view of the effective reality, i.e. of the things whenever they happen, regardless of desirability (Foucault 2007).

Therefore, whereas discipline pretends to regulate everything (things should not be released to their own destiny), security mechanisms let things be. Freedom, though not absolute, is essential to security because the constant emergence of new differences is indeed security's raw material. Even though details are essential for disciplinary mechanisms, differences must be restricted. As Dillon and Lobo-Guerrero (2008) have pointed out, "The function of security was to rely on details that were not intrinsically good or bad but necessary; inevitable, as with so-called natural processes" (Dillon and Lobo-Guerrero 2008: 281).

Consequently, security is what actually structures the social-technical order of telecare, but not in terms of expelling any uncertainty or possibility of danger. On the contrary, security is a way of bringing uncertainty into the production of order. It is something that includes and addresses uncertainty, and which at the same times extends it productively. Consequently, telecare is a security system because it manages events as risks and tries to modulate the user's life and habitat according to them rather than because it protects the user's habitat from some sort of threat.

A Securized Milieu

The notion of security conveys a shift that was barely visible in the panoptic apparatuses and sheds light on the spatiality of telecare and other similar services. According to Foucault (2007), while discipline produces places, and places within places (homes, offices, regions and territories), security works through *milieus*. Discipline is mainly centripetal; it works by segmenting spaces. Its aim is to build up spaces where power operates boundlessly and unwaveringly; thereby, visibility is total and any difference is perfectly localized. On the contrary, security apparatuses are constantly extending and multiplying their limits. They are centrifugal. They are incessantly articulating new elements. We have seen this in telecare, i.e. the provision of safety depends on the capacity to connect and coordinate more and more remote resources in real time. The telecare service must be well articulated with different kinds of institutional agents: ambulances, police, GPs, outpatient departments, etc. but it must also transform relatives, neighbors, informal caregivers, and any others that would be able to give a hand, into a manageable resource for telecare services. This capacity for open-ended articulation is essential to put the user in the centre of the social and healthcare network and to make the teleoperator some sort of conductor who steers this network to respond to the specific situation of the user. Thus, providing safety consists not only in the production of stable and consistent articulations, but relies on the transformation of these articulations.

This study of telecare has brought to light a number of differences between security and disciplinary spaces. Whereas discipline always operates by hollowing out the space and in empty spaces that must be totally organized by it, security apparatuses make up and modulate a milieu. Discipline, as Foucault (2007) explains, props up places (a *total institution*, a home, a city, a state, etc.) because there is an aspiration of total accomplishment. Each of these disciplinary spaces is a world in itself, and a world containing other worlds. On the contrary, security spaces must be understood as milieus, spaces modulated by events.

> Security will try to plan a milieu in terms of events or series of events or possible elements, of series that will have to be regulated within a multivalent and transformable framework. The specific space of security refers then to a series of possible events; it refers to the temporal and the uncertain, which have to be inserted within a given space. The space in which a series of uncertain elements unfold is, I think, roughly what one can call the milieu. (Foucault 2007: 35)

Thus, in contrast with the idea of place and the very idea of network, the notion of milieu suggested by Foucault enables us to explain how security is 'eventfully' spatialized. Both places and networks are spaces that emerge as the output of a constant work of excluding what appears to be uncertain. What is at stake in these spaces, therefore, is the inscription of a specific order in different sorts of materialities, whether they might be mobile or static, such us bodies or architectures. In this sense, places and networks could be considered disciplinary spaces. In contrast, milieus are defined as eventful spaces and, hence, constantly modulated by practices and techniques that try to deal with events in order to include them productively (see Deleuze 1992).

For example, installing telecare devices does not necessarily entail setting up clear boundaries between safe and unsafe areas. On the contrary, the home space is defined by multiple uses of the telecare device that overflow those boundaries. Therefore, telecare's aim is not to totally transform the home in order to perfectly adjust the functions of the service to the spatial and practical configuration of the house, but instead to set up the house and uses of the telecare based on possible events. The instructions given to the users by the telecare service are a clear instance of this. These are a few examples taken from the fieldwork: 'As you may likely fall down, it is extremely important that you do not take the pendant off when taking a shower; and even if you do want to take it off, leave it on the floor in order to be able to reach the button just in case you do fall'. 'Because hard of hearing is a common problem among the elderly, the telecare terminal is not being installed next to the TV because the volume is normally turned up and teleoperators might not be able to hear what is going on in the house'.

Thus telecare does not operate according to what should be going on in a defined territory, but instead operates according to what potentially might be happening in an open and thus changing milieu.

Our telecare study reveals the absence of a utopian project. There is not a clear and defined ideal of the good telecared house because there is no expectation of perfection and harmony. The aim is simply to optimize the effects of living with a telecare system, reduce what might be 'bad' and maximize what might be 'good' in a specific situation. Thus, it seems that a norm without morality is operating in this case. The related norm is operating technically from the point of view of the effective reality. Hence the reason why the production of a user identity perfectly adjusted to the service is also not present in the telecare service. The service is not pretending to construct the perfect user. Telecare as a security mechanism is polyfunctional because there is no attempt to fix a specific use for each thing. Instead, telecare's aim is to manage the possible effects that an increasing multiplicity of uses might have for the service and for the daily live of the user. Or, to put it in other words, its main duty is to adjust the different milieus[8] in order to get an over-all effect: an increase in the quality of life optimizing the cost-benefit balance in care delivering. This entails slightly changing the daily routines of the user and the spatial configuration of the house in order to obtain an increase in the user's perception of auto-control, to better cope with a worsening of any problem linked to the ageing process, and to maximize the success possibilities of an urgent intervention conducted by the telecare operators. Therefore, as discussed elsewhere (López and Domènech 2009), the tension between an autonomy embodied in a *body-at-risk* and an autonomy embodied in a *vigorous-body* (that is mediating the appropriation process of the service by the users and explaining why some users are reluctant to press the alarm) is not an obstacle in itself. It is indeed quite the opposite. We are considering telecare as a security mechanism because it deals with different sorts of autonomies that are trying to combine and optimize (not neutralize) the possible effects that each has on health, wellbeing, economic cost, technological use, etc.

However, stating that telecare is a security mechanism does not necessarily mean that there is no discipline anymore. As we have seen, different forms of body training and spatial transformations of places are also involved in telecare operation. These practices and techniques are operating according to a different logic, however. Instead of building up a coherent and stable order, their main purpose is to modulate or adjust, based on a constant estimation of potential events and its effects, any element involved in telecare. This is what defines a security mechanism and enables us to think about the specificities of our case, moving beyond the traditional discourses on community care and deinstitutionalization.

Final Remarks

The concept of security space as milieu enables us to move a step further away from the hyperbolic interpretations of the proliferation of technologically advanced surveillance systems. Power is not as extended as ever before just because it has fully invaded outdoor space, and neither are we facing a new logic of power due to the proliferation of ICTs and the impact of the rapidity of information transmission

---

8   We have coined the concept of habitality to analyze how these milieus enact multiple forms of being at home with telecare (see López and Sánchez-Criado 2009).

in different social areas. Thus, the dismantling of total institutions and the boom of community care at home, as well as the development and implementation of telecare services addressed to more and more specific needs, are not what is at stake. According to this study on telecare, the most striking conclusion is that the logic of security might be affecting virtual care environments, such as telecare, as much as home care services or nursing houses. Fundamentally, what we are trying to say is that the increasing importance of telecare and e-health means that more and more contexts, whether enclosed or open, are configured as security spaces because they are configured as a response to possible events (harmful or beneficial). We might conclude by saying that being outdoors or *transmural* is not the most salient aspect of the community care fostered by telecare technologies. What defines this community care is that spaces of care are increasingly going to be shaped by security practices that address the riskiness of everyday life.

References

Akrich, M. and B. Latour. 1992. A Summary of a Convenient Vocabulary for the Semiotics of Human and Nonhuman Assemblies, in *Shaping Technology/ Building Society*, edited by W. E. Bijker and J. Law. Cambridge: MIT Press, 259–308.

Bashshur, R. L., T. G. Reardon and G. W. Shannon. 2000. Telemedicine: A New Health Care Delivery System. *Annual Review of Public Health*, 21: 613–637.

Bogard, W. 2006. Surveillance assemblages and lines of flight, in *Theorizing surveillance: the panopticon and beyond*, edited by D. Lyon. Cullompton, Devon: Willan Publishing, 97–122.

Castel, R. 1991. From dangerousness to risk, in *The Foucault effect: Studies in Governmentality*, edited by G. Burchell, et al. Chicago: The University Chicago Press, 281–298.

Cooper, R. 1992. Formal Organization as Representation: Remote Control, Displacement and Abbreviation, in *Rethinking Organization. New Directions in Organization Theory and Analysis*, edited by M. I. Reed and M. Hughes. London; New York: Sage Publications, 255–272.

Dean, M. 1999. *Governmentality: power and rule in modern society*. London; Thousand Oaks, Calif.: Sage Publications.

Deleuze, G. 1992. Postscript on the Societies of Control. *October*, 59 IS: 3–7.

Dillon, M. and L. Lobo-Guerrero. 2008. Biopolitics of security in the 21st century: an introduction. *Review of International Studies*, 34: 265–292.

Donzelot, J. 1979. *The Policing of Families*. 1st American Edition. New York: Pantheon Books.

Dubet, F. 2002. *Le Déclin de l'Institution*. Paris: Seuil.

Editorial. 2002. Ancianos en soledad. *El País*, 24 August: 8.

Essén, A. 2008. The two facets of electronic care surveillance: An exploration of the views of older people who live with monitoring devices. *Social Science & Medicine*, 67(1): 128–136.

Ewald, F. 1991. Insurence and risk, in *The Foucault effect: Studies in Governmentality*, edited by G. Burchell, et al. Chicago: The University Chicago Press, 197–210.

Foucault, M. 1977. *Discipline and Punish*. New York: Pantheon.

Foucault, M. 2007. *Security, territory, population: lectures at the Collège de France, 1977–1978*. Houndmills, Basingstoke, Hampshire; New York: Palgrave Macmillan.

Gordon, C. 1991. Governmental rationality: an introduction, in *The Foucault effect: Studies in Governmentality*, edited by G. Burchell, et al. Chicago: The University Chicago Press, 1–51.

Graham, S. 1998. Spaces of surveillant simulation: new technologies, digital representations, and material geographies. *Environment and Planning D: Society and Space*, 16(4): 483–504.

Graham, S. and D. Wood. 2003. Digitizing Surveillance: Categorization, Space, Inequality. *Critical Social Policy*, 23(2): 227–248.

Haggerty, K.D. 2006. Tear down the walls: on demolishing the panopticon, in *Theorizing surveillance : the panopticon and beyond*, edited by D. Lyon. Cullompton, Devon: Willan Publishing, 23–45.

Haggerty, K. D. and R. V. Ericson. 2001. The surveillant assemblage. *The British Journal of Sociology*, 51(4): 605–622.

Hine, N., A. Judson, S. Ashraf, J. Arnott, A. Sixsmith, S. Brown and P. Garner. 2005. Modelling the Behaviour of Elderly People as a Means of Monitoring Well Being. *User Modeling 2005*: 241–250.

Leith, K. H. 2006. "Home is where the heart is… or is it?" A phenomenological exploration of the meaning of home for older women in congregate housing. *Journal of Aging Studies*, 20(4): 317–333.

Lianos, M. 2003. Social control after Foucault. *Surveillance & Society*, 1(3): 412–430.

López, D. 2008. Aproximación a la topología de la Teoría del Actor-Red. Análisis de las espacialidades de un servicio de Teleasistencia Domiciliaria in Tecnogénesis. La construcción técnica de las ecologías humanas, edited by T. Sánchez-Criado. Madrid: AIBR, 113–137.

López, D. and M. Domènech. 2008. On inscriptions and ex-inscriptions: the production of immediacy in a home telecare service. *Environment and Planning D*, 26: 663–675.

López, D. and M. Domènech. 2009. Embodying autonomy in a Home Telecare Service. *Sociological Review*, 56(s2): 181–195.

López, D. and T. Sánchez-Criado. 2009. Dwelling the Telecare Home: Place, Location and Habitality. *Space and Culture*, 12(3): 343–358.

Lyon, D. 1994. *The Electronic Eye: The Rise of Surveillence Society*. Minneapolis: University of Minnesota Press.

Lyon, D. 2006. *Theorizing surveillance: the panopticon and beyond.* Cullompton, Devon: Willan Publishing.

May, C., T. Finch, F. S. Mair and M. Mort. 2005. Towards a wireless patient: Chronic illness, scarce care and technological innovation in the United Kingdom. *Social Science & Medicine*, 61(7): 1485–1494.

Miller, J. and R. T. Miller. 1987. Jeremy Bentham's Panoptic Device. *October*, 41 IS: 3–29.

Mira Johri, F. B. 2003. International experiments in integrated care for the elderly: a synthesis of the evidence. *International Journal of Geriatric Psychiatry*, 18(3): 222–235.

Mort, M., T. Finch and C. May. 2009. Making and Unmaking Telepatients. Identity and Governance in New Health Technologies. *Science, Technology & Human Values*, 34(1): 9–33.

Nettleton, S. 2004. The Emergence of E-Scaped Medicine? *Sociology*, 38(4): 661–679.

Percival, J. and J. Hanson. 2006. Big brother or brave new world? Telecare and its implications for older people's independence and social inclusion. *Critical Social Policy*, 26(4): 888–909.

Rose, N. 1996a. *Inventing our selves: psychology, power, and personhood.* Cambridge, England ; New York: Cambridge University Press.

Rose, N. 1996b. The Death of the Social? Re-Figuring the Territory of Government. *Economy and Society*, 25(3): 327–356.

Rose, N. S. 1999. *Powers of freedom : reframing political thought.* Cambridge, United Kingdom; New York, NY: Cambridge University Press.

Winthereik, B. R. and S. Vikkelsø. 2005. ICT and Integrated Care: Some Dilemmas of Standardising Inter-Organisational Communication. *Computer Supported Cooperative Work (CSCW)*, 14(1): 43–67.

Chapter 3

Exploring the Affordances of Telecare-related Technologies in the Home

Chris Tweed

Introduction

Tang et al. (2000: v) define telecare as: '... the delivery of health and social care to individuals within the home or wider community outside formal institutional settings, with the support of devices enabled by information and communication technologies.' In this broad definition, the technologies range from automated reminders and medicine dispensers, to environmental monitoring, centralised door and access control systems. In spite of the diversity of telecare systems and devices, most can be analysed as consisting of one or more of the following components: sensors, agents, actuators and networks. Most health providers in the UK have devised procedures—the Single Assessment Process in England, the Unified Assessment Process in Wales, and the Single Source Assessment in Scotland—for deciding if a person would benefit from a telecare installation in her home. These are person-centred methods of assessing needs and risks, taking into consideration the physical and psychological health of the patient as well as social context. As such, they are practical tools for deciding if telecare is likely to benefit an individual with particular needs in a given social context. There is less emphasis on the impact telecare may have on the way people experience their homes or the changes it may bring to spatial practices within the home. Even for seemingly straightforward encounters between people and devices there are other facets to explore if we are to appreciate the role technology assumes in our lives (McCarthy and Wright 2004).

The impact of telecare on the way people use their homes should be of particular interest to architects and building designers and yet this is rarely discussed in the literature, which tends to focus on the immediate interaction between the user and specific devices. Hence, there is a need for better understanding of how telecare technologies are likely to impinge on the spatial practices people follow as part of everyday living in the home. The aim of this chapter is to examine existing theories about how people experience and use space (mainly in the home) and then to consider how these might be applied to studying the impact of telecare on spatial practices. Two main areas are considered—phenomenology and ecological psychology—and the chapter looks at the links between ideas that have originated from each of these areas: principally, lived space and affordances.

The chapter begins by reviewing some of the theories describing people's experience of space in architecture. Many of these theories originate in the work of phenomenologists of the last century, such as Maurice Merleau-Ponty (1962). Merleau-Ponty's work is particularly relevant to discussions about telecare since much of it focuses on the role of the active body in experiencing the world. This is followed by a discussion of the concept of affordance and its subsequent appropriation by Donald Norman (1988) to become a dominant theme in discussions about the usability of products and computer software. This discussion is widened to consider how ecological psychology and phenomenology can work together to provide useful insights into affordances within specific cultural settings, before looking at previous applications of affordance to architecture. Finally, the chapter considers how these theories relate to previous studies of what people value at home and the way in which telecare systems and services may disrupt, or enhance, home life.

Phenomenological Treatments of Space

Some of the most detailed descriptions of our experiences in and around buildings are found in the writings of 20th Century phenomenologists. The most celebrated of these are Bachelard's *Poetics of Space* (1964) and Heidegger's essays, "Building Dwelling Thinking," "...Poetically Man Dwells..." and "The Origin of the Work of Art" (Heidegger 1993). Otto Friedrich Bollnow has also addressed experience of the built environment, but his influence on the architectural community in English-speaking countries has been hampered by the fact that there is no translation of his major work, *Mensch und Raum* (Bollnow 1963) from German into English. The only widely available summary of his thought has been provided by Nold Egenter (2009). From what is known through Egenter's translation and interpretation, Bollnow could be a major contributor to debates about space in architecture, but it looks as though we will have to wait for a full translation for this promise to be fulfilled. Architectural theorists have drawn heavily on Heidegger and his architectural interpreters, principally Norberg-Schulz (1963) who is responsible for much of the interest in phenomenology in architecture. Elaboration of these themes continues in the work of phenomenologists such as Edward Casey (1998) and David Seamon (1993). However, it is the work of Merleau-Ponty (1962), particularly because of his emphasis on the body as the locus of perception, who offers the most promising analysis of experiencing the spatial world, which is attracting the attention of many contemporary architectural theorists.

For Merleau-Ponty, there is a fundamental difference between *lived space* and the objective space of the geometer. Lived space, is space as experienced by an individual as opposed to objective space, which is assumed to be independent of specific individuals. Objective space is described using standard units of measurement along the three Cartesian coordinates. It is the space architects usually work with. The conventional tools of the architect's trade encapsulate objective space—scale rules, CAD (Computer Aided Design) systems, etc. Most phenomenologists would argue that objective space is imposed on lived space to meld it to fit standardised scales of measurement. Objective space is also known as 'third person' space.

Merleau-Ponty's references to lived space are often about the experiences of individuals suffering some form of disability or mental illness. These cases accentuate the fact that people can experience the world very differently, but should not be taken to suggest that differences exist only between 'normal' people and others who are considered to be in some way lacking or deficient in their physical, mental or reasoning abilities. Merleau-Ponty's phenomenology is detailed and complicated and so it is impossible to do it justice here. Suffice to say that it emphasises the role of the body as the locus for experiencing the world and the importance of activity (and intentionality) in how we perceive our surroundings. Perception of the world is underpinned by the ever-present revealing of possibilities of acting in the world, the so-called 'I can.' To perceive, therefore, is often to perceive what it is possible for me to do. He notes that the body develops an intuitive sense of how to position itself optimally in relation its environment and the objects within it, as in the viewing of a picture in a gallery: 'For each object, as for each picture ..., there is an optimum distance from which it requires to be seen, a direction viewed from which it vouchsafes most of itself: at shorter or greater distance we have merely a perception blurred through excess or deficiency.' (Merleau-Ponty 1962: 348)

Merleau-Ponty's influence on architecture, therefore, has been to emphasise the role of the body and kinaesthetics in our experience of space, including architectural space. He claims that our everyday interaction with buildings is not as cerebral as many theorists would have us believe, and that there exists a level of bodily experience that is prior to our analysis of architecture. But even Merleau-Ponty's architecturally indifferent description of the perception of space implies a lived-body which is assumed to be healthy if not positively athletic. Although he chooses extreme cases of disability to advance his argument, these are always contrasted with an implied healthy exemplar. Once again possible experience is narrowed, in this case by assumptions about the type of bodies we might have.

A key element of Merleau-Ponty's theory of spatial perceptions concerns the adjustments individuals make following changes to the shape and size of the body when equipped with spatial extensions, such as the blind man's cane or a woman wearing a large feather in her hat. In addressing the changes in body perception, Merleau-Ponty's work is relevant to concerns that surface in discussions about the efficacy of telecare and the changes it brings to the perception and experience of space. Toombs, for example, cites Merleau-Ponty's examples of how the body can be extended through the use of objects that through habitual use effectively become part of the body:

> Merleau-Ponty notes that in the normal course of events (through the performance of various habitual tasks) the embodied individual incorporates objects into

bodily space. For example, the woman who habitually wears a hat with a long feather intuitively allows for the extension of the feather when she goes through a doorway. The experienced typist no longer views the keys of the typewriter as objective locations at which she must aim. (Toombs 1993: 82)

Furthermore, Toombs recognises that 'illness causes a constriction in the lived spatiality of the patient in that the range of possible actions becomes severely circumscribed' (Toombs 1993: 82). In this respect, Merleau-Ponty's phenomenology sets the scene for further studies of the impact of telecare on spatiality, but its focus is firmly on the perception of space from the individual's point of view. To understand the impact of telecare, it is necessary to consider the issues from a broader perspective.

Hodological Space

A small number of architectural theorists have promoted the treatment of space as habitat for humans. Nold Egenter (Egenter 2009) devotes much of his research towards developing these ideas, though much of his thinking has yet to enter the architectural mainstream. One of Egenter's primary sources is Otto Bollnow's book, *Mensch und Raum.* As noted above, this work is only available in German and so very little is known about the detailed arguments it presents. Egenter has provided a helpful summary, but the work is rarely referenced elsewhere.

One of the more interesting ideas to emerge from Egenter's summary of Bollnow's work is hodological space. Egenter translates Bollnow definition as: '[Hodological space] is a type of space which differs absolutely from mathematical space. Path-space or hodological space, corresponds to the factual human experience during movement between two different points on a map. It is absolutely different from the geometrical line which connects two points.' (Egenter 2009)

Hodological space, therefore, is structured by paths and points, more precisely 'fixed' or 'zero' points, from which we organise our world at home and (temporarily) in hotel rooms and other resting places. Hodological space is replete with social and cultural significance. To use Egenter's example, he highlights the huge psychological distance between two points in adjacent apartments that are geometrically very close (30 to 40 cm) but separated by a wall, pointing out that architects rarely appreciate 'what somebody goes through in term of physical and psychological stress, to go from one of these points to the other.' As for Merleau-Ponty then, Bollnow's conception of space is very different from the geometer's, which is the normal way of treating space in mainstream architectural design, for example in computer aided design (CAD) systems.

There are obvious parallels in the ideas promoted by Bollnow and the body-centred perception of Merleau-Ponty. Of particular interest to the current topic, is the observation that people organise space to suit their practical and psychological desires at any fixed point. One can see this casually in meetings, cafés and other social gatherings. People establish 'camps' in these settings and then proceed to structure the space around them by developing an orientation to the rest of the social group, marking out territory through the placement of objects and setting up an 'efficient' layout of 'tools' required for the immediate purposes (eating, note-taking, reading etc.). If the same place is used intermittently over consecutive periods, one can observe the 'camping' behaviour of people returning to the same seats and locations within a space over repeated visits. People refer to 'my seat' and can be noticeably disturbed if someone else occupies what they consider to be their place. This phenomenon is even more evident at home.

The Idea of Affordance

The idea of affordance is the invention of the psychologist, J. J. Gibson, who developed it as part of his broader ecological approach to visual perception:

> The affordances of the environment are what it offers the animal, what it provides or furnishes, either for good or ill. The verb to afford is found in the dictionary, but the noun affordance is not. I have made it up. I mean by it something that refers to both the environment and the animal in a way that no existing term does. It implies the complementarity of the animal and the environment.... (Gibson 1979: 127)

The key element of an affordance, therefore, is that it neither belongs solely to the environment nor to the organism, but is the result of their encounter. Such a view undermines the objective treatment of function in architecture in which it is assumed that buildings and their components have definite and usually single-valued functions that occupants need to adhere to if the building is to be used 'properly.' Unfortunately, the inevitable failure resulting from a mismatch between users and buildings is often blamed on the inability or unwillingness of users to adapt their practices to the requirements of the design. Not surprisingly this has nurtured a view of architects as being arrogant and out of touch with their clients. The concept of affordance is potentially liberating for designers because it reveals the myriad possible meanings the products of design can have for different people, or organisms.

Where an affordance differs from what might be called a straightforward property of the environment is that it depends on the interaction between the environment and an organism. As Gibson states:

> An affordance is neither an objective property nor a subjective property. It is both. An affordance cuts across the subjective-objective dichotomy and in doing so highlights the inadequacy of this dualistic thinking. It is equally a fact of the environment and a fact of nature. It is both physical and psychical, yet neither.

An affordance points both ways, to the environment and to the observer. (Gibson 1979: 129)

Gibson recognised the need for organisms to possess certain abilities to perceive and make use of affordances, but he stopped short of providing a detailed account of human-environment relations and, in particular, how affordances vary from one person to the next or even for the same person at different times. Refinements to the definition of affordance have been developed by Turvey (1992) and Chemero (Chemero 2003).

An important corollary of affordance, therefore, is the complementary idea of *effectivity*, which refers to the abilities of an organism. Thus, for example, one can talk about body effectivities as the ability of a given organism to realise an affordance because of its body scale relative to features of an environment. The possible affordance of something being available or accessible to a person may depend on the effectivity of reach, if the object is placed some distance away, for example, on a high shelf (Rochat et al. 1999). In the context of telecare, effectivities are hugely importance since the effectivities of users are likely to change rapidly over time with the result that the affordances offered by an environment (such as a telecare-equipped dwelling) will also change rapidly. Similarly, the effectivities of a person equipped with telecare devices will change as new devices are introduced and their operation is mastered. Telecare installations have the ability to change both the properties of the environment and the effectivities of people inhabiting the environment. In a sense, an artefact, by offering an affordance to a suitably skilled person, becomes a new effectivity. To use Merleau-Ponty's example, a blind person's cane initially affords obstacle-sensing and in the hand of the blind person, it enhances their effectivities to include (say) 'able to detect solid obstacles within one metre.'

This combination of affordances and effectivities would appear to offer a powerful conceptual tool for discussing the impact of telecare, or indeed any technological intervention in an environment, on the possibilities it offers different users. When applied to design, the ideas of affordance and effectivities should draw attention to the importance of pairing effectivities with properties of artefacts to realise affordances. The match between environment and effectivities is not spatially uniform, which gives rise to areas within an environment that are particularly suited to the effectivities of different organisms. These areas are known as ecological niches, and are an important part of Gibson's ecological psychology, though as Smith and Varzi (Smith and Varzi 1999) note they are not treated in a particularly rigorous fashion. The concept of a niche for individual organisms is potentially powerful in exploring how people use and settle in places. There is also a correspondence between the niche and Bollnow's fixed points discussed above.

Affordances in Design

The design of computer interfaces was one of the first areas to explore the application of Gibson's idea of affordance, closely followed by the design of products and devices. Donald Norman pioneered the use of affordance in his book on human interaction with designed artefacts, *The Psychology of Everyday Things*, or *POET*[1] Norman's use of affordance is controversial and departs from Gibson's original definition in two important ways (Koutamanis 2006): first, *POET* focuses mainly (though not exclusively) on perceived affordances rather than possible affordances, which is what interested Gibson; and second, *POET* is mainly concerned with man-made artefacts and how they communicate their intended operation to potential users, which is in contrast to Gibson's interest in natural environments. Norman's examples include the design of telephones, and door handles. For the latter, Norman makes the point that the relatively simple operation of opening a door usually requires a decision on the part of a user whether to push or to pull but that the information imparted by the design of many doors often fails to convey the correct message about this choice. For Norman, therefore, affordance primarily implies perceived affordance and the task for designers is to ensure that users perceive the intended function of an artefact. This is the fundamental difference between Gibson and Norman, and indeed Norman has claimed that if he were to revise the text of *POET* again he would replace every occurrence of 'affordance' with 'perceived affordance.'

Regardless of whether one agrees with Norman's appropriation of Gibson's idea, there is no doubt that he has succeeded in raising the profile of affordance and particularly its potential in the field of design. Without this, it is doubtful the insights it offers would have been realised.

Affordances and Architectural Design

Current interest in applying the idea of affordances to architecture is not new, but the past approaches have tended to follow Norman's lead rather than Gibson's broader focus (Tweed 2001). The approach adopted in *POET* is developed further in recent research into affordances in engineering design by Maier et al. (2009). Although the declared target of their research is architecture, the provenance of their method lies in engineering and their approach can be considered as highly rational. They combine the idea of affordance with other methods for assessing the fulfilment of function—such as Function Task Interaction (FTI) (Galvao and Sato 2005) developed to investigate product architecture—and incorporate ideas from Kim et al. (2007) on the assessment of the usability of environments. They introduce an expanded set of relations, which includes human-human and artifact-

---

1   Later editions have been published under the title, *The Design of Everyday Things*, or *DOET*.

artifact affordances, in addition to the Gibsonian human-artifact description of affordance. Human-human affordances presumably refers to the observation that people are able to offer each other services, but this is far removed from the Gibsonian focus on relations between organisms and their environments. Similarly, artifact-artifact affordances seem to refer to the functional relationship between building elements, such as wall providing support to a roof. There is a possible logic for making these connections—it encourages us to think beyond the assumed functional relationship—but in doing so it dilutes the fundamental concept of affordance.

This enthusiasm for the general interpretation of affordance as a co-dependency between things and the proliferation of different types of affordance leads to a general confusion about what is meant by affordance in architecture (Maier et al. 2009). Indeed it becomes clear that the authors associate affordances with specific parts of a building, as in: 'Windows afford the transmission of light, and hence illumination of the interior environment as well as a view of the exterior environment. Operable windows may also afford the exchange of air, and in extreme cases even defenestration.' (Maier et al. 2009: 396); and again, 'it is not merely sufficient to design an artefact that possesses certain affordances' (Maier et al. 2009: 405). In the discussion of the affordances of a window, an occupant is implied, but her effectivities are not considered. As noted above, one of the most useful aspects of affordance is that it makes us think about effectivities and what is needed on the part of an organism to realise a possible affordance in the environment. A window, therefore, will only allow the exchange of air if it can be opened, and that action will require a person with certain capabilities as well as a window in a certain location with a usable opening device, etc.. This definition, therefore, misses the crucial pairing of environment and organism that Gibson made central to his definition of affordance.

Kim et al. (2007) provide a more restrained application of the concept of affordance, which they use to evaluate the usability of different types of spaces. Their study consists of four main phases: space and user studies; affordance investigation; interaction exploration; and, benchmarking simulations. The authors use this method to analyse the affordances of conference rooms, starting with a specific case study. The analysis follows the stages outlined above and produces lists of objects and building components in the room and the possible interactions with these. The authors distinguish between high- and low-level affordances, with high-level affordances for a conference room including 'enter/exit-ability, prepare-ability, present-ability, discuss-ability, and conclude-ability' (Kim et al. 2007: 7). Thus, they arrive at a detailed checklist of the desired affordances of a conference room. Perhaps because this is a more modest application, its utility is more obvious. However, the matrix approach is unable to capture differences in people's effectivities which, as we have seen above, is central to a well developed notion of affordances in an environment. In fact, there is no mention of this aspect of affordance. Their approach is essentially a functional analysis of the objects and building components in a conference room. A similar approach could be applied to assessing the impact of telecare installations, but it would miss important aspects of people's interaction with telecare in the home mainly because it is so focused on functional ability and neglects people's broader experience of place.

Koutamanis (2006) highlights the differences between spaces and objects as targets for an analysis of affordances. Spaces, in contrast to products, lack the 'handy interfaces' that form part of solid objects. Spaces are less tangible and serve as containers of activity rather than being involved directly. Designers often intuitively grasp the possibilities for action in different types and sizes of spaces, in the manner that Bachelard's phenomenology would suggest. However, standard architects' reference sources containing ergonomic data ignore the subtle meanings implicit in their depictions of spatial layouts and requirements, such as having to walk sideways after closing the door of a WC cubicle. Ergonomic data embedded in floor plan and sectional diagrams inevitably create a false sense of spatial security, particularly to the fledgling designer, in so far as it is easy to assume that if a design meets the minimum space standards it will provide a satisfactory environment. This is precisely the kind of thinking Alexander criticised in the development of his pattern language (Alexander et al. 1977), which owes much to the notion of affordance.

There is a hubristic trap waiting for the architect in all this, as Koutamanis acknowledges: 'The built environment is generally background to human goals and actions—rarely the subject itself.' (Koutamanis 2006: 358). It is therefore wise to remember that architecture is only one part of the environment and on any given day, for any given person, may contribute more or less to her experience. Still, it is impossible to deny that the design of a building and its systems, including telecare, determines many of the possible activities and the quality of their realisation.

The Problem with Affordances

The shortcomings of various theoretical developments surrounding affordances are not limited to attempts to apply it to architecture, though many of the problems of application in that domain probably result from the lack of grounding in the more scientifically oriented field of psychology. Martin Oliver (2005) provides a trenchant critique of affordance that demands a serious response if we are to continue to use the idea for important work in the future. His main objections are:

- Gibson's commitment to direct perception does not allow any room for learning or for cultural influences on perception.
- The definition of the relationship between subject and object is never made clear.

He dismisses the notion that the affordances that exist between an organism and its environment is the set of all possible interactions, because these are unspecifiable and therefore infinite. In this case, affordances are speculative rather than analytic. Affordances are possibilities, but it is not possible to specify them all in advance, since for any given organism and its environment the possibilities are probably limitless. In other words, there are probably an infinite number of things I can do with a chair, some of which will be more feasible or obvious than others. What makes them feasible is the ease with which an affordance can be realised and what makes them obvious is more aligned with cultural norms. Gibson made no allowance for the influence of culture in the perception of affordances, and ecological psychologists are divided on whether and how this can be incorporated in an extended theory (Chemero 2003). Others, outside the field, perhaps naively, happily graft cultural influences on to the theory of affordances without showing much concern, as we shall see in the next section.

On this evidence the opportunities for a coherent theory of affordances in architecture seem limited, in spite of Gibson's plea for precisely that type of application: '… a glass wall affords seeing through but not walking through, whereas a cloth curtain affords going through but not seeing through. Architects and designers know such facts, but they lack a theory of affordances to encompass them in a system.' (Gibson 1979). And yet, it is difficult to abandon the latent promise of such a theory even if it may eventually have to be supplemented by other insights, such as those derived from phenomenology.

Affordances and Phenomenology

A link between affordances and phenomenology can be found in a paper by Dreyfus which discusses the current relevance of Merleau-Ponty's thought to theories of embodiment (Dreyfus 1996). An affordance is seen as a culturally conditioned response to an artefact, such as a chair. Dreyfus argues that a chair affords sitting on because we have the kind of bodies that bend at the back of the knee and because sitting is a culturally defined norm in many, but not all, situations. The relations between anatomical possibilities of the human body and culturally sanctioned behaviour are complex. There are, for example, situations where it is not acceptable for individuals to sit. Furthermore, there are situations in which a person refuses an invitation to sit down because of the expected behaviour and power relations of the social context. Nonetheless, the notion of affordance would seem to offer a useful starting point from which to consider what buildings can offer different users.

Dreyfus' explanation of affordance is edifying:

> J. J. Gibson, like Merleau-Ponty, sees that characteristics of the human world, e.g. what affords walking on, squeezing through, reaching, etc. are correlative with our bodily capacities and acquired skills, but he then goes on, in one of his papers, to add that mail boxes afford mailing letters. This kind of affordance calls attention to a third aspect of embodiment. Affords-mailing-letters is clearly

not a cross-cultural phenomenon based solely on body structure, nor a body structure plus a skill all normal human beings acquire. It is an affordance that comes from experience with mail boxes and the acquisition of letter-mailing skills. The cultural world is thus also correlative with our body; this time with our acquired cultural skills. (Dreyfus 1992)

For Dreyfus, then, skills are key effectivities over and above the capabilities of an organism's body scale, physical strength, dexterity and perceptual acuity.

In summary, the idea of affordance continues to attract attention from ecological psychologists, but also from designers and architects. The latter are drawn to it largely by Norman's appropriation in *POET*, which he acknowledges is mainly about perceived rather than actual affordances. The application of the idea varies widely from Gibson's vague treatment of relations between organisms and their environments to the precise, functionalist approach within the engineering tradition. Despite its documented shortcomings and logical inconsistencies the work on affordance does offer three concepts: first, it emphasises organisms and environments as co-creators of affordances—an affordance can only emerge from the interaction between an organism and an environment; and second, it is clear that an affordance can only emerge if the organism has the capabilities (body scale, strength, skills) needed to realise it—a step is only a step if it can be climbed; and finally, there is a cultural brake on what is possible as an affordance, so while an affordance may be there, it can be ruled out by social or cultural norms.

Spatial Organisation and Practices in the Home

Literature about the home is voluminous. It includes architectural treatises on the history of housing and technology in dwellings (Rybczynski 1988), sociological analyses of comfort and hygiene (Shove 2003), the meaning of objects in the home (Csikszentmihalyi and Rochberg-Halton 1981), everyday practices surrounding the use and maintenance of the fabric (Shove et al. 2007), through to philosophical accounts of what it means to be at home in a fundamental sense and independently of any physical enclosure (Steinbock 1995). For the purposes of this study, our point of departure is the definition offered by Mary Douglas of home as 'a kind of space' (Douglas 1991), a view that is embraced by others seeking to understand the notion of home in relation to care (Schillmeier and Heinlein 2009).

Douglas goes on to characterise the home as 'a pattern of regular doings' (1991: 287) noting that 'home starts by bringing some space under control' (1991: 289). This and related anthropological and sociological accounts underline the complexity of the home and the rituals, power relations and feelings it contains. In many of these studies, the role of the architecture is secondary, and it is possible to imagine that the same analyses might be relevant in a different physical setting. Is the architecture relevant? Clearly it is, on one level, since the layout of spaces, their size and the environmental conditions they offer (for good or for ill) influence (if not determine) the activities that can take place within them. Beyond that, there are indications that less tangible characteristics of the home are important to how people experience their homes in ways they would not in a hotel, to invoke Douglas' example. Much of this is carried by the furniture and possessions we fill our homes with (Csikszentmihalyi and Rochberg-Halton 1981), but even a bare house can evoke memories of past experiences, perhaps through remembered past thoughts and feelings when contemplating the view from a window.

Every home organises space by allowing its inhabitants to orient themselves to what is not home, usually the outside world. Even temporary homes establish an orientation, and it can be argued that people continuously create 'homes' in a variety of different spatial and social settings beyond their residences as they 'dwell' in places such as offices, schools, cinemas and even when colonising space on the beach. This orientation towards (and separation from) the world defines the front and backstage areas that are fundamental to our sense of security and privacy (Goffman 1959). The need to have a place where one can relax and be oneself seems critical to a sense of well-being.

Studies of spatial configuration reveal significant differences between how different social groups organise and use space in the home. In a study of lifestyles in small terraced houses in residential areas of inner London, Hanson found that traditional working class households tend to favour a highly structured home space with rigid allocation of activities to enclosed spaces (Hanson 1998). In their relation to the street, the interiors of traditional working class homes were concealed from the street by placing objects in ground floor windows, whereas new middle class homes frequently removed all obstructions from the front windows and also tended to use minimal coverage of window (blinds) late in the evening, such that the illuminated interior was clearly visible to passers-by late into the evening. For traditional working class households the back door was frequently left open or on a latch but in new middle class homes the back door was often removed altogether and burglar alarms and formal door furniture were fitted to the front door, which became the formal entrance. Hanson's study, though conducted in the 1970s, reminds us that we should not make generalisations about homes, assuming that most people will have similar attitudes. Indeed the study found that for many aspects the new middle class home was an inversion of the traditional working class home: 'each way of configuring space embodied the dominant cultural practices of a different socio-economic group.' (Hanson 1998). This is borne out by the author's own studies of home settings.

Affordances of the Home Environment

The affordances of interior space differ from the affordances of objects, mainly because space must cater for multiple users at once if it is to support social interaction (Kim et al. 2007). In theory it should be possible to apply the function-oriented methods of analysis developed by Kim et al. for conference rooms to exhaustively identify the affordances of each space within a dwelling. However, as noted previously, in practice it is impossible to list all possible affordances of an organism-environment pair, as they are probably infinite, especially if they were to take on board the range of effectivities of all possible occupants. Activities at home are far less restricted and goal-oriented than for a conference room, which makes the task of enumerating affordances harder still. This section therefore aims to pinpoint those affordances that are the defining characteristics of home and not just buildings in general.

As noted previously, existing attempts to apply the idea of affordance to buildings has tended to veer towards a highly functional analysis of the programmatic requirements of architectural space and in doing so neglect specifically architectural qualities of spaces such as their adjacencies and form. But the aesthetic experience of designed spaces is important and extends beyond conversations among the architectural cognoscenti to embrace the emotions people have in response to their configuration of home. In a study of families living in public housing projects in the Netherlands, Paul Pennartz (2006) focused on the experience of 'atmosphere' at home, which he labelled as the most comprehensive characteristic of a place. The study divided the themes emerging from interviews into two groups: sociopsychological factors and architectural features of the home and their influence on the experience of atmosphere. Pennartz found that the most important sociopsychological factors were: communicating with each other; being accessible to one another; being relaxed after having finished work; being able to do what one wants to; and being occupied, or absence of boredom. The architectural themes contributing to the experience of atmosphere were: the arrangement of and connections between the rooms; the size of the rooms; and the shape of the rooms. The study concludes that material and architectural properties constitute the potential environment, but that for pleasantness to be realised requires a connection between these and the conceptual level of the sociopsychological factors.

The home acts as a container for the activities of its occupants. These include exercising those associated with personal care such as eating, washing, dressing, sleeping, socialising and engaging in those activities that might be loosely labelled as 'living'—the activities that keep boredom at bay and stimulate the mind; communicating with others both inside and beyond the boundaries of the home; and those that make life interesting and worth living. These latter activities are missing from the functional analyses of homes—telecare-equipped or otherwise—described above but are often crucial to establishing a good quality of life.

Although it is impossible to define all potential affordances for every individual inhabiting a given home environment, it is possible to establish some high-level affordances. For example, it seems reasonable to assume that every home should afford comfort (thermal, visual, aural and physical), privacy, security, shelter from the climate, accessibility and configurability. It is worth remembering that in creating these positive affordances, it is inevitable that they may also be seen as negative affordances at some times for some people. The secure 'four walls' of a dwelling can serve as a fortress, a cocoon but equally as a prison or a firetrap.

As for any building, a dwelling needs to offer the conditions needed to support occupants' activities. Hence it needs to provide comfort and safety as well as the space needed to perform a range of varied activities. Beyond this, as Douglas notes, home must provide space that allows occupants to store provisions to respond to future needs. For Douglas, this is the single most distinguishing feature of home, and it is contrasted to the lack of storage provided in a hotel. However, there are other important defining characteristics for home, such as the privacy it creates—its backstage quality—which allows occupants to be themselves largely free from the cultural norms that dictate social behaviour when in company. It is doubtful this would qualify as an affordance under any of the definitions provided above, but it is essential to the constitution of home.

Within the home there is a hierarchy of spatial organisation which is in part derived from the affordances offered by different locations. In Gibsonian terms, these locations can be viewed as niches that emerge from a good match between organism (occupant) and environment (the building or the objects it contains). The classic example of a home niche is a favourite chair, often defined by its affordance of television viewability, constrained by distance, absence of glare, angle of incidence to the screen, etc. In homes with multiple occupancy, each occupant is likely to establish different niches to inhabit depending on relations with other occupants, their roles within the home and immediate needs. Such niches achieve semi-permanent status through habitual occupation of spaces and a process of negotiation involving all of the occupants or habitual users of the home. A chair becomes 'my chair', just as readily as a room becomes 'my room.' Home, because it is familiar, allows us to establish habitual behaviours, many of which are tied to specific rooms and even locations within those rooms. The connections between activities and places become sedimented and routine to occupants, but may be invisible to visitors.

So, how might telecare installations change the affordances of the home? Is it possible to apply such an idea to understanding how people perceive and make use of their homes? To answer these questions we first need to clarify what an installation might consist of.

Telecare Systems

The range of services, systems and appliances that fall under the general rubric of telecare is vast and growing. Not surprisingly many of the techniques and hardware originate in computer science, artificial intelligence and biomedical engineering. Most telecare systems usually consist of one or more of the following basic technologies, which are often combined in a single device:

- sensors—sense and measure specific variables indicating the state of the environment or person, e.g. environmental sensors (for temperature, humidity, smoke and CO2), movement sensors, pressure sensors, telemedicine systems for measuring pulse rate, blood pressure and other vital signs;

- agents—decide what action (if any) to take based on information provided by sensors and internal logic, the 'intelligence' of the system, but may be little more than a bimetal switch;
- actuators—devices that do something when instructed by an agent, including alarms, displays, lights, or communications devices, such as telephones and intercoms.

In addition to these components, there are infrastructural elements such as communications networks that operate within the home—intercom, CCTV systems for monitoring entrances—and links to care service providers outside the home—pendant alarms, customised telephones, panic buttons. There is also usually some form of hub device which links the various components together and often provides the system intelligence.

Telecare, therefore, embraces a vast range of services, systems and devices. It is impossible to consider all of these in this chapter. The discussion below offers a more general consideration of how these technologies might impact on spatial practices in the home.

Telecare and Affordances in the Home Environment

There are established applications of phenomenology and the theory of affordances in the domain of telecare. Dreyfus, for example, enlists the support of prominent phenomenologists, especially Merleau-Ponty, in his critique of the claims emanating from artificial intelligence, which is a key component of the agency used in the more advanced telecare systems (Dreyfus 1992). And Norman's affordance-based observations on product design are clearly relevant to many of the devices used in telecare installations, from LCD displays to panic buttons or any item of telecare equipment that relies on interaction with a user (Norman 1988). As interesting as these debates are, most of the issues they raise are well documented, but not particularly relevant to spatial practices, though they may exert some influence on them. To study the possible impact of telecare on spatial practices in the home suggests a different tack, one which draws on the phenomenological insight of the body as the locus of experience and the variation of the availability of affordances across a landscape. This latter aspect of Gibson's original theory of affordance and organism-environment interaction is neglected in discussions about the affordances of architecture and interiors, but is central in, for example, Ingold's discussions about landscape and skills (Ingold 2000).

There is a convergence of Gibson's description of niches in the environment, Steinbock's phenomenological description of territory and terrain (Steinbock 1995), and Bollnow's ideas about fixed points and paths in hodological space. These ideas demand serious consideration outside the scope of this chapter. However, it will be useful to sketch how these ideas might be used to explain the impact of telecare, as one type of technology, within the home environment. The remainder of this section seeks to identify ways in which these ideas might be applied.

In previous research, the author recorded the location of televisions and other contemporary appliances in different homes in an effort to discover how these organise space in the home. This was part of study into the impact of technology on the configuration of the home environment. Through this it became obvious that televisions, computers and telephones can have a major influence on how people organise space. Familial homes, owned by older people, showed traditional arrangements of seating still focused on the redundant fireplace with newer technologies, such as the television, placed in subordinate (but accessible) positions causing minimal disruption to existing spatial configurations. The fireplace often contained a display of bulrushes or other low maintenance natural symbols. This type of arrangement would continue to favour conversation as entertainment rather than viewing television. In other cases, notably among students, new appliances often completely dominated the furniture arrangements. In one example, a desk with computer, monitor, keyboard and printer was placed in front of an disused fireplace with an attractive, decorative surround, thus obscuring the major architectural feature of the space. Clearly, for some, functionality easily trumps aesthetics.

Spatial configurations based on television viewing requirements are probably the easiest to decipher in the modern home, because of the need for line of sight, avoidance of screen glare, etc. For all members of an audience to be able to see the screen requires that they keep a certain distance from the screen and within a view cone, which provides acceptable angles of vision. Within the limits of the viewing distance and the view cone, some positions will be better than others and this is when hierarchies within a social group may come into play, with more dominant members occupying the prime positions. Values for viewing distance and angle will depend on the size of the screen, what is being displayed on the screen and on the visual and auditory perceptual abilities of the viewers. For many, it will be impossible to read characters (such as *Teletext*) on a given size of screen beyond a critical distance and at a certain angle. Spatial arrangements based on viewing televisual images are unlikely to provide optimal positions for reading text from the screen in shared viewing situations. It can be concluded that the practices of television viewing are replete with spatial significance, and that individuals will position themselves optimally relative to the screen, just as Merleau-Ponty suggests we do when viewing paintings in an art gallery.

The home environment, therefore, can be viewed as a mini-landscape in which organisms (occupants) with different effectivities find their niches according to the affordances these home-places offer. The inhabiting of such niches, however, is underdetermined by physical capabilities and will be influenced by social and cultural characteristics of the setting and the individuals. The home is an infinitely more complex landscape than it first appears. Similarly, the home can be viewed as hodological space with numerous fixed points lying on the paths that connect rooms, cupboards, seats, appliances, beds, sinks, toilet, shower, doors and windows.

Introducing new technologies invariably alters existing patterns of space usage. Telecare adds new fixed points and nodes to the existing paths of hodological space. These are not just new points to visit along a path, as they take on new meanings related to the activities they afford. The placing of an intercom terminal acquires associations with modifying the security of the dwelling in an entirely different way to answering the door. And this may only affect one occupant and not others. Of course the installation of a remote door opening system represents a substantive change to the boundary dwelling allowing others to enter on the basis of a conversation with occupants, and possibly unobserved. When coupled with surveillance and monitoring equipment installed in the home and other keyholders, the boundary of the home effectively dissolves and the distinction between front and backstage no longer exists (Percival and Hanson 2006). However, we have also noted changes to paths and fixed points in the home with far less intrusive telecare interventions: once occupants learn the consequences of triggering sensors they can deliberate avoid or seek to fool sensors to achieve a desired action from the system. People are surprisingly adept at learning how to exploit the system to reach their goals. Sensors may be obscured, disconnected and reoriented.

This short exploration of the application of phenomenology and affordances to telecare suggests new areas to develop not just to understand the impact of telecare but to identify better how people inhabit and orient themselves within a technologically sophisticated landscape. The final section suggests possible directions for future research.

Conclusion

This chapter has sought to explore the possibilities of bringing an alternative conception of architectural space to bear on the issues that arise from introducing telecare to the home. The chapter has drawn on the phenomenological treatment of space, the development of J. J. Gibson's ecological psychology and particularly its concept of affordance, as well as subsequent exegesis of affordance by Norman in product design and previous applications of this body of work to architecture and buildings. The main premise of this work is that space is not uniform or homogenous. The Cartesian description of space is of minimal use when we consider how people make use of and experience lived space.

In examining different appropriations of Gibson's idea of affordance we discover wide variation with some approaches amounting to little more than a functional analysis of space. While such analyses have their uses, they sacrifice much of the richness of our everyday encounters with designed spaces and thereby miss opportunities for enhancing our interaction with telecare systems. It is perhaps not surprising that busy healthcare professionals focus on the practical needs of their charges while also considering the wider social setting and the impact of technology on this. In one sense, this chapter is arguing for the icing on the cake, and if there are problems with telecare they are at the more extreme end of sophisticated artificial intelligence applications, rather in the day-to-day use of panic alarms and environmental sensors. Assigning agency to systems rather than people raises issues to be discussed elsewhere.

Existing applications of affordance are found to be lacking when applied to the home. Although there are obvious advantages in applying Norman's version of affordance to the design of individual telecare components, there are no precedents for applying Gibson's treatment of landscape to enclosed spaces. This seems to be an area ripe for development. As Smith and Varzi (1999) note, the ecological niche has a potentially wide application in many disciplines, which has yet to be realised. However, Smith also suggests that Gibson's treatment of the niche is not well developed and seeks more rigorous treatment in Husserl. Steinbock is likely to be helpful in progressing this research, with his account of territory and terrain in Husserl's phenomenology (Steinbock 1995).

It is also reasonable to suggest that to be useful to analysing people's use of space, the idea of affordance must be stretched to embrace feelings as well as physical activity. The functional affordances of, for example, exit-ability, configurability and so on need to be supplemented by private-ness, and other more phenomenological descriptors. The fact that a home affords a sense of security if it is capable of preventing unwelcome entry from others under the control of the occupant is important. It will also only afford feelings of safety if it can allow quick and easy egress for the occupant in the event of a fire. However, we need to be vigilant against extended notion of affordance getting so diluted it is no longer recognisable.

Finally, there are many possible criticisms that can justifiably be levelled at the motives underlying increasing use of telecare, particularly if it is used as a substitute for human carers. However, it is important to remember that telecare and assistive technology allow many people to stay at home, sometimes with families who could not provide enough care otherwise, rather than enter an institutional setting. For many, an altered home environment is still the best place to be.

References

Alexander, C., Ishikawa, S. and Silverstein, M. 1977. *A Pattern Language: Towns Buildings Construction*. New York, Oxford University Press.

Bachelard, G. 1964. *The Poetics of Space: the classic look at how we experience intimate places*. 1994 edition. Boston, Beacon Press.

Bollnow, O. F. 1963. *Mensch und Raum*. 2nd edition., W. Kolhammer.

Casey, E. 1998. *The Fate of Place: A Philosophical History*. Berkeley and Los Angeles, University of California Press.

Chemero, A. 2003. An Outline of a Theory of Affordances. *Ecological Psychology*. 15(2): 181–195.

Csikszentmihalyi, M. and Rochberg-Halton, E. 1981. *The Meaning of Things: Domestic Symbols and the Self*. Cambridge, Cambridge University Press.

Douglas, M. 1991. The Idea of a Home: A Kind of Space. *Social Research.* 58(1): 287–307.

Dreyfus, H. L. 1992. *What computers still can't do: a critique of artificial intelligence.* Revised edition. Cambridge, Mass.; London, MIT Press.

Dreyfus, H. L. 1996. The Current Relevance of Merleau-Ponty's Phenomenology of Embodiment. *Electronic Journal of Analytic Philosophy.* 8.

Egenter, N. 2009. *IMPLOSION.* [Online] Available at: http://home.worldcom.ch/negenter/ [accessed: October 2009].

Egenter, N. 2009. *Otto Friedrich Bollnow's Anthropological Concept of Space.* [Online] Available at: http://tinyurl.com/yjk725c [accessed: October 2009].

Galvao, A. B. and Sato, K. (2005). Affordances in product architecture: Linking technical functions and users' tasks. *ASME 2005 International Design Engineering Technical Conferences & Computers and Information in Engineering Conferenc.* Long Beach, CA, ASME: 11.

Gibson, J. J. 1979. *The Ecological Approach to Visual Perception.* Boston, Houghton Mifflin.

Goffman, E. 1959. *The Presentation of the Self in Everyday Life.* Anchor.

Hanson, J. 1998. *Decoding Homes and Houses.* Cambridge University Press.

Heidegger, M. 1993. Building Dwelling Thinking, in *Basic Writings: Martin Heidegger*, edited by. D. F. Krell. London, Routledge.

Ingold, T. 2000. *The Perception of the Environment: Essays in livelihood, dwelling and skill.* London, Routledge.

Kim, Y. S., Kim, M. K., Lee, S. W., Lee, C. S., Lee, C. H. and Lim, J. S. (2007). Affordances in Interior Design: A Case Study of Affordances in Interior Design of Conference Room Using Enhanced Function and Task Interaction. *ASME 2007 International Design Engineering Technical Conference & Computers and Information in Engineering Conference IDETC/CIE 2007.* Las Vegas, ASME: 10.

Koutamanis, A. 2006. Buildings and Affordances, in *Design Computing and Cognition '06*, edited by. J. S. Gero. Dordrecht, Springer: 345–364.

Maier, J. R. A., Fadel, G. M. and Battisto, D. G. 2009. An affordance-based approach to architectural theory, design and practice. *Design Studies.* 30: 393–414.

McCarthy, J. and Wright, P. 2004. *Technology as Experience.* Cambridge MA, The MIT Press.

Merleau-Ponty, M. 1962. *The Phenomenology of Perception.* Routledge and Kegan Paul Ltd.

Norberg-Schulz, C. 1963. *Intentions in Architecture.* London, Allen & Unwin Ltd.

Norman, D. A. 1988. *The Psychology of Everyday Things.* New York, Basic Books.

Pennartz, P. J. J. 2006. Home: The Experience of Atmsophere, in *At Home: An Anthropology of Domestic Space*, edited by. I. Cieraad. Syracuse, Syracuse University Press: 95–106.

Percival, J. and Hanson, J. 2006. Big brother or brave new world? Telecare and its implications for older people's independence and social inclusion. *Critical Social Policy.*

Rochat, P., Goubet, N. and Senders, S. 1999. To reach or not to reach? Perception of body effectivities by young infants. *Infant and Child Development.* 8: 129–148.

Rybczynski, W. 1988. *Home: A Short History of an Idea.* London, Heinneman.

Schillmeier, M. and Heinlein, M. 2009. Moving Homes: From House to Nursing Home and the (Un-)Canniness of Being at Home. *Space and Culture.* 12(2): 218–231.

Seamon, D. 1993. *Dwelling, Seeing and Designing: Towards a Phenomenological Ecology.* Albany, SUNY Press.

Shove, E. 2003. *Comfort, Cleanliness and Convenience: The Social Organization of Normality.* Berg Publishers.

Shove, E., Watson, M., Hand, M. and Ingram, J. 2007. *The Design of Everyday Life.* Oxford, Berg.

Smith, B. and Varzi, A. C. 1999. The Niche. *Noûs.* 33(2): 214–238.

Steinbock, A. J. 1995. *Home and Beyond: Generative Phenomenology after Husserl.* Evanston, Illinois, Northwestern University Press.

Tang, P., Curry, R. and Gann, D. 2000. *Telecare: new ideas for care and support @ home.* Bristol, The Policy Press.

Toombs, S. K. 1993. *The Meaning of Illness: A Phenomenological Account of the Different Perspectives of Physician and Patient.* Dordrecth, Kluwer.

Turvey, M. 1992. Affordances and prospective control: An outline of the ontology. *Ecological Psychology.* 4: 173–187.

Tweed, C. 2001. Highlighting the affordances of designs: mutual realities and vicarious environments, in *Computer Aided Architectural Design Futures 2001*, edited by. B. d. Vries, J. v. Leeuwen and H. Achten. Dordrecht, Kluwer Academic: 681–696.

Chapter 4

Clutter Moves in Old Age Homecare

Peter A. Lutz

Introduction

As a noun, *clutter* is defined as a disordered state or heap of objects while as a verb it is to move about in a bustling manner or to strew or amass things in a disorderly way (*Collins English Dictionary* 2003). This interweaving of cluttered things and movements is an appropriate starting point for the present chapter. By figuring *clutter moves* in old age homecare the aim is to engender a modest contribution to a growing curiosity in movement and old age homecare. It asks what array of socio-technical implications emerge in old age homecare ecologies when analytical attention focuses on how older people and cluttered homely things move together.

Clutter may irritate norms about order, productivity and cleanliness and is often considered a cultural taboo (cf. Douglas 1966; Douglas and Wildavsky 1983). An Internet search of the term *clutter* helps illustrate this point. The most popular phrases include *clean the clutter* with about 400,000 hits and *control the clutter* with over 600,000 hits. However, *remove the clutter* achieves the highest score with roughly 2,200,000 hits. Here *removal* clearly emerges as one of clutter's most widespread associations. More generally, this demonstrates how clutter is inherently linked, through idiomatic expressions, with the value of *clutter-free living* and *clutter removal* tactics.[1]

The inclination towards clutter removal is no less apparent in United States old age homecare ecologies. Certified programmes that offer training in home modification for *Aging-in-Place*, a popular US label that denotes the constellation of infrastructures and processes designed to support home living, is a case in point. Two clear examples are the *Certified Ageing-in-Place Specialist* (CAPS) programme of the *National Association of Home Builders* (NAHB) and the *Senior Move Management Training* (SMMart) programme of the *National Association of Senior Move Managers* (NASMM). A central

---

1 Numerous popular self-help books with instructive strategies for removing home clutter provide further illustration. A few self-explanatory titles will suffice: *The Clutter-Busting Handbook: Clean It Up, Clear It Out, and Keep Your Life Clutter-Free* (Emmett 2005) and *Houseworks: Cut the Clutter, Speed Your Cleaning, and Calm the Chaos* (Ewer 2006) or *Organize It!: How to Declutter Every Nook and Cranny In and Outside Your Home* (Kaufman 2006). The list goes on.

element of these programmes is learning how to recognize, neutralize, as well as prevent so called 'environmental risks' for older people at home including the removal of their clutter.

Various articles about how to manage home clutter appear regularly in the bi-monthly magazine of the *American Association of Retired Persons* (AARP). In one of these articles, Dudley (2007) proffers that the do-it-yourself clutter removal trend is related to the emergence of the professional organizer in the United States. Some examples are the *National Association of Professional Organizers* (NAPO) and support groups like *Clutterers Anonymous* (CLA) or *Messies Anonymous* (MA). These aspire to help arm individuals in their own private wars on clutter. Several *reality television* programmes like *Clean Sweep, Clean House* or *Extreme Makeover Home Edition* evidence this general trend. These programmes often depict how a person's clutter is whisked away overnight and replaced with a cover story look-a-like from the latest home design magazine. This apparent obsession of decluttering in the United States helps to further contextualize this chapter.

The ethnographic stories related here aim to reveal how removing home clutter is not always easy or appreciated among older people. In one case an elderly gentleman of 94 found it extremely difficult to rid his home of cluttered memorabilia. None of his four children had much interest in his family things. Simultaneously his nature held him back from just, in his words, '*bagging it up and throwing it all away*'. Several participants spoke openly about their clutter and even referred to it in derogatory terms as '*old stuff*' or '*junk heaps*'. Yet on closer analysis these cluttered things often emerged as meaningful and ordered. As this chapter will show, the entangled movements of people and their home clutter are closely linked with home ecologies of old age care.

The heuristic *clutter moves* is offered as a means to rethink the relationship between older people, their things and old age homecare including its technologies. Part of this potential stems from the multiple definitions encapsulated in the term *movement*. According to the *Oxford English Dictionary* (OED), three key definitions are: a change of place or position; a mental impulse especially one of desire or aversion; and an organization, coalition or alliance of people working to advance a shared political, social or artistic objective (*Oxford English Dictionary* 1992). Hence movement contains both literal (e.g. *motion*) and figurative (e.g. *emotion*) dimensions. Incidentally, the OED explains that the term *emotion* is derived from the Latin prefix *e-* meaning *out* and the verb *movere* meaning *to move* producing the definition, the action of moving out (*Oxford English Dictionary* 1992). With this nuance in mind, the *act of moving* or *removing* becomes an *emotional* move per definition. This is a curious contradiction with the more contemporary usage which depicts emotion as an internal mental or conscious state of feeling. Yet it also underscores how movement is thoroughly entangled with *material-semiotic* (Haraway 1997) relationships.

The broader exploration of movement has enjoyed increasing popularity in scholarly pursuits. For instance, Urry has recently coined the phrase *mobility turn* defined as a 'post-disciplinary [...] way of thinking through the character of economic, social and political relationships' (Urry 2007: 6). Then, in human-computer interaction (HCI), there is a long-standing interest in mobile devices and systems. This includes *assistive technologies* for elderly and disabled people. Meanwhile, movement has played a central role in studies of science, technology and society (STS). Latour's influential concept *immutable mobiles* is one example (Latour 1987). Another is Mol and Law's refiguring of actor-networks as *fluids* and *flows* (Mol and Law 1994). Anthropology is no exception. Here the *mobility turn* originates in studies of migration, space and identity as well as territorial boundaries (cf. Jansen and Löfving 2009). In the book *Friction*, for example, Tsing employs movement's literal (mobility) and figurative (social mobilization) dimensions to launch an analysis of the Indonesian environmental movement (Tsing 2005). In relation to old age homecare one recent example is provided by Schillmeier and Heinlein who explore various implications of moving from one domestic residence to another in old age (Schillmeier and Heinlein 2009).

So what about technology? The literal or *standard view* frames it as a broad category of material artefacts, systems, techniques and machines much apart from the social sphere. Pfaffenberger for one recognizes that both popular and scientific accounts generally adopt this standard view (Pfaffenberger 1992). This contrasts with the *symmetrical view* of technology as fundamentally socio-technical relationships consisting of both human and nonhuman actors (cf. Latour 1987, 1999). This latter view increasingly retained in anthropology and STS is also assumed here. The heuristic of *clutter moves* is an attempt to contribute a symmetrical analysis of *people-things* in old age homecare ecologies. Hence this does not presuppose a clear distinction between the literal and figurative or Haraway's *material-semiotic* in figuring the ways people and their cluttered things entangle one another.

The term *ecology* is also employed throughout this text. It is used to encompass the multiple spatial-temporal entanglements of heterogeneous entities including people and their cluttered things. The OED credits the zoologist Haeckel with coining the term in the late 1800s (*Oxford English Dictionary* 1992). He derived the term from the Greek words *oikos* meaning *house* and *logos* meaning *science*. Hence the notion of heterogeneous home ecologies seems within Haeckel's original intentions for the term where home and old age care practices merge. Meanwhile the phrase *old age homecare* is employed to fold together multiple *ageings*, *homes* and *cares*. The intention here is not to black-box the multiple practices bracketed by this phrase. Rather it is offered as abbreviated shorthand while the movements of people and things are traced through these heterogeneous ecologies as clutter moves.

The ethnographic fieldwork on which this chapter is based employed semi-structured interviews, participant-observations and home visits recorded with a handheld video camera over a four-month period. Fieldwork engagements were scheduled on a one-to-one basis. The home interviews involved approximately twenty older people. With participant consent, these interviews incorporated a walk-through tour of their domestic surroundings. They often lived alone but when available their paid and unpaid caregivers, including family members, were also asked to participate. More casual follow-up meetings and participant-observations transpired with roughly a half-dozen of the older participants as well as their caregivers.[2]

Next, a brief review of a familiar move in old age homecare studies is presented: falling older people. It suggests that the gerontological and geriatric literature tends to oversimplify clutter as a hazard leading to the risk of falling. The subsequent sections question such depictions from an ethnographic view focused on other ways people and clutter move around in old age homecare ecologies.

Scientific Literature on Falling Older People

This section reviews a series of gerontology and geriatric studies of old age homecare that deal with the risk of falling. As noted this literature typically identifies home clutter as a major hazard in domestic environments. The *Centers for Disease Control and Prevention* (CDC) offers one of the most compelling statements found. It reports that falls are the leading cause of injury deaths among people over the age of sixty-five and home clutter is noted as a significant environmental factor implicated in one third of these falls (CDC 2007: 29). The remainder of this section elaborates how similar research, primarily from the US and UK, has figured such cluttered movements.

Several studies stress how clutter threatens independent living and physical movement at home. For example, Wilson and Rodgers who, upon reviewing research of physical activity and exercise among older people with cardiovascular disease, confirm that cluttered homes are unsafe for exercise and lead to falls and injuries (Wilson and Rodgers 2006: 635). Then, in a British study Wherton and Monk argue that the independence of older people with dementia can be supported by minimizing their home clutter (Wherton and Monk 2008: 2). This link between independence and hazardous clutter is also echoed by Börsch-Supan et al. who stress that falling is a major threat to the ability of maintaining one's household and independence in old age (Börsch-Supan et al. 2005: 45).

Another tendency is the factoring of home clutter as a hazard in the development of various risk assessment schemes. For instance, Messecar et al. in their study of elderly home caregivers' *environmental modification strategies* reference a *home modification intervention* scheme that designates the removal of home clutter (Messecar et al. 2002: 358). In another study, Evans and Kantrowitz

---

2   The *American Anthropological Association's Code of Ethics* guided all participant engagements including informed consent. Participant pseudonyms are used throughout this text to protect personal identities.

use a *housing composite scale* that relies in part on respondent's assessments of *cleanliness/clutter* to measure *environmental risk exposure* (Evans and Kantrowitz 2002: 310).

The overwhelming majority of these studies employ quantitative analysis of rated questionnaire responses from older people as well as care professionals. Here the home clutter category is often calculated alongside multiple other hazards. For example Gershon et al. invited 738 Resident Nurses (RNs) to rate a series of hazards in all out-patients' households, not only older persons. According to these RN respondents, more than a third of the patients' households contained *environmental hazards* including home clutter (Gershon et al. 2008: 6). This study used *messy home/clutter* as one of ten *environmental and physical hazards* in the development of their *household hazards construct*. These were measured in relation to *biological, chemical* and *violence-related hazards* (Gershon et al. 2008: 5). Presumably the threat of home clutter warrants comparison to such things as domestic violence and poisonous chemicals. In a similar study that explored reasons for falling, Zecevic et al. utilized a telephone survey to collect the views of seniors and health care providers (Zecevic et al. 2006). A total of 28 categories were measured and included, from highest to lowest frequency: *Balance, Weather, Inattention, Medical conditions, Indoor obstacles (including clutter), Surface hazards outside, Slip-trip-stumble, Dizziness, Attitude, Muscle weakness, Vision, Footwear, Motor control, Medications, Old age, Hurry, Stairs, General poor health, Alcohol, Indoors hazardous surfaces, Assistive devices, Inactivity, Chance event, Frailty, Lighting, Poor service, Unfamiliar environment,* and *Fear of falling* (Zecevic et al. 2006: 374). Again, home clutter is mixed together with an entire series of possible hazards that increase the risk of falling.

Unfortunately, the distinctions of these categories such as the difference between *slip-trip-stumble* and *chance event* are highly ambiguous. With some imagination it is feasible that clutter could be implicated in a few or several of its sister categories, depending on how each respondent perceives clutter. While a correlation between home clutter as hazard and the risk of falling is well documented, there is a surprising lack of description about what the category of home clutter actually entails or why it is there in the first place. Here are a few exceptions.

In their interview study with care professionals and older people, Blythe et al. quote participant(s) verbatim on home clutter: '[There is] *too much furniture in the home.* [Occupational therapists] *hate those rugs.* [Older people] *always trip over them.* [...] *They leave yards and yards of cable and they tend to pull it wherever they go. They get up to go to the loo, answer the phone and they fall over the cable*' (Blythe et al. 2005: 676). With reference to such irritated statements and combined with a panel on falls prevention, these authors prescribe that environmental assessments of home should occur in coordination with older persons' discharge from the hospital (Blythe et al. 2005: 676). Another quick glance is offered by Moylan and Binder who cite cluttered walkways in relation to community occupational therapy services with programmes designed to assess and modify home environmental risks (Moylan and Binder 2007: 496). Loose rugs and cords are also mentioned by Russel et al. in a statistical study of observed or measured risk factors for falling (Russel et al. 2006: 1091).

While items such as rugs or telephone cables are explicated in some of these studies the category of home clutter is generally left relatively unpacked, thereby masking the diversity of objects and situations. Here a disjuncture emerges between recommendations for clutter removal and knowledge about the multiple forms and meanings that it may occupy. One plausible explanation is that home clutter is only one factor in a multi-factored risk analysis of falls and thus left undistinguished. However, if moving around with clutter is such a pervasive concern then one must assume there are multiple practices at work to produce its emergence.

The quantitative methodology predominantly employed to sample care professionals' and older persons' own risk ratings in such studies may also be implicated in the interpretive gap between the category and the contents it enfolds. Presumably, participant responses were collected after fall incidents for statistical analyses. It is therefore feasible that the research categories or terminologies used were unable to fully catch the nuance of interpretations as they occur in practice. In other words, the ranking terms employed may end up cloaking other more subtle processes. Hence, *home clutter* and *hazard* or *risk* in one context may mean something very different in another and thereby go unnoticed in the final analysis. This in turn may summon the inaccurate portrayal of home ecologies and the various ways people and their clutter relate within them.

Such queries are acknowledged in an article by Marshall et al. who opt for the method of home visits. This was instead of telephone surveys to gain a more qualitative assessment of falls in relation to the home and household clutter (Marshall et al. 2005: 99). A second example is found in Zecevic et al. who qualitatively link clutter with the risk of falling (Zecevic et al. 2009). This latter article assesses how the adoption of a case method used in the Canadian transportation industry. This was to investigate safety and accidents could be adopted for the systematic analysis of falling older people. Their research is built around a series of cases framed to identify systematic causes. One case in particular focuses on home clutter as hazard and how it contributed to a fall incident:

> Case 17: Trip Over the Box. – At approximately 11:00 p.m., an 84-year-old woman fell in her apartment kitchen. She tripped over a heavy box that, earlier that evening, she filled up with sweets and jars of jam. Earlier that week, she was instructed to remove all sweet foods from home, after being diagnosed with Type 2 diabetes. She left the box on the floor near the kitchen entrance. She planned to move the box out of the way in the morning and to give it to her daughter the following weekend. Habitually, the senior leaves items on the floor when she plans to take them out of the apartment. This creates clutter that was present at several places in her bedroom during the investigation. The senior had cleaned out her closet and prepared multiple bags of items for her daughter to take away. On the day of the fall, it was the first time the senior left an obstacle (box) in the middle of the walking path to the kitchen. After finishing her before-

bed routine in the bathroom, she turned off all the lights in the apartment (to conserve energy) and quickly walked from the bathroom to the kitchen to throw a piece of wrapping paper into the garbage bin. Her right foot hit the heavy box, both feet blocked behind the box; she lost balance and fell forward hitting the carpeted floor with her face. The impact broke her nose; her glasses made deep indentations around her eyes that later caused severe facial bruising. [...] At the time of the fall, the senior lived alone (Zecevic et al. 2009: 690–691).

This passage is among the more descriptive accounts of home clutter in the gerontology/geriatric literature focused on fall risks among older people. It makes clear that home clutter as hazard entails much more than rugs carelessly strewn across the floor in concert with a mess of telephone lines. To the contrary, home clutter as hazard may also take the form of carefully organized and contained boxes or bags marked for removal. Equally important is the degree of explicitness revealed in the above case when compared with the prevailing quantitative risk assessments. Almost by accident, it anticipates the broader ethnographic exploration of clutter moves and how it interweaves older people and their cluttered collections. For instance, it implicates the healthcare professional's directives to remove sweet foods, in turn prompted by a diagnosis of type 2 diabetes. It also connects the practice of giving things away that is inherently entangled with family relations while living alone in old age.

Such analyses begin to push clutter moves beyond the association with threatened independence or risk assessments which factor clutter simply as a hazard; sometimes even in tandem with poisonous household chemicals or domestic violence. While it is not my intention here to deny the risk of falling, I do suggest that there are other useful *cluttered* avenues of analysis to move down. Qualitative descriptions such as the one above suggest a fruitful alternative. The next section takes up this challenge. The heuristic of clutter moves is engaged to survey a series of ethnographic stories which highlight ways cluttered things entangle themselves with old age homecare ecologies.

Collecting and Distributing

The scientific literature in gerontology reviewed above has more often than not factored clutter and movement as potentially a perilous relationship. Physically moving around in home areas, cluttered or otherwise, is clearly a challenge for ageing human bodies. Yet, what other stories surface besides risky clutter when ethnographic attention is paid to how older people and their clutter moves? This section builds on evidence from ethnographic fieldwork to propose other kinds of clutter moves. In part, this focuses on how clutter is *collected* within the home while also *distributed* and entangled within ecologies of old age homecare.

Collecting and Distributing Cluttered Family Ties

The earlier example of a woman tripping and falling over sweet foods she had collected in a box to give to her daughter suggests that clutter is not simply scattered about the home in careless fashion. My ethnographic research also supports this paradoxical assertion: home clutter generally surfaces as ordered collections (see also Makovicky 2007; Swan et al. 2008). Nevertheless, this was not without frustration. Here the ebbs and flows of clutter moves are explored in terms of how they permeate home ecologies including family relations.

Meet Patrick. He was a delightful and intelligent 94-year-old widower. His wife had passed away seven years earlier but he still lived in the ranch-style house he had designed and built in the 1950s. He proudly showed me a picture taken from an airplane some years before and displayed in the entrance way. This birds-eye view documented a car in the driveway and a sailboat that he had built as well as a field of Christmas trees ready for harvest. Since the picture was taken he had sold the car and the trees, and had given the boat to his son in California. He had christened the boat *'Misadventure'* and kept its stamped lifebuoy as a token. His other three children, all daughters, lived within a 15-mile radius. Although close in terms of travel distance, they were all too busy to spend as much time with him as he wished.

Patrick explained that in the interim between breakfast and lunch he often read, listened to radio news, did light housework or worked on his *'cleaning the clutter'* project. This was *'old stuff'* that he and his family had accumulated over the years. It included newspaper clippings, family correspondence, pictures, clothing, decorations and other miscellaneous objects of some sentimental value. One could say memorabilia filled his house like thick molasses. He reflected on his saving nature and linked it with growing up in the Depression years when resources were scarce. He explained that it would not be so difficult to sort through it if he only knew what to do with it all. He had difficulty getting verification from his main recipients: his children. He had even asked his paid care worker to help him label and sort things into manila file folders. However, due to his increasingly poor eyesight it was difficult for him to read the labels she had written. He realized that some of his organizational problems could have been solved with a computer. But since he had never learned to use one he did not intend to start now. During one of my visits Patrick directed me to a stone fireplace hearth in the living room where he had arranged and labeled four shoeboxes, one for each of his children.

Like the woman's boxes of sweet food, Patrick's shoeboxes exemplify how clutter moves between inward concentration and outward distribution. However, this was not careless removal but rather a highly selective process seeped in meaning. Resourcefulness and attempts to save things perceived valuable, no matter how insignificant, imply an economic dimension to clutter overlooked in recommendations for clutter removal surveyed earlier. Material-temporal fluctuations were also observed in several cases where people had accumulated special things that needed a new home. Among my participants I found collections of old cameras, typewriters, computers, family picture albums, dolls and stamps to piles of yarn, shelves of quilting fabrics and even a porcelain frog collection numbering over two hundred figures. These were prepared for giving but *not* throwing away. Yet, as Patrick noted, distributing clutter was often hampered with frustration. For instance, when I asked another 87-year-old female participant if it was difficult to remove her clutter she exclaimed, *'I am gradually working on it. I try to give it away. I mean the kids and people have given me stuff over the years and I've got to put it somewhere.'* The intended receivers are not always interested.

Patrick's shoeboxes illustrate collecting-distributing clutter moves that connect family ties. Such moves also intersect and redefine his home ecology. For instance, what once was a fire hearth is now also a mailroom; what once was an eating area is now primarily a sorting and cleaning the clutter workroom. Indications of risk seem far removed. Meanwhile, clutter moves engaged with self-care imply another form of collecting-distributing. *Nesting selves* is a key practice that emerges here.

Collecting and Distributing Cluttered Nesting Selves

Arleen, a 76-year-old widow, is a prime example of how clutter enables self-nesting practices. She was an articulate and eccentric woman who lived alone in a two-floor semi-rural house that was increasingly proving difficult for her to maintain. She suffered from a number of chronic ailments, including type 2 diabetes. She drove a little orange and rusty 1970s Toyota pickup and walked around with the help of two antique shooting sticks used as canes. These are exemplary of her ingenuity. She explained that the shooting sticks had been purchased at *Harrods* in London many years earlier. However, when she decided she required walking assistance she began employing them as canes. She was proud that she had managed to salvage them for a 'proper' use. She also explained that they made her feel more ominous and had accidentally stunned a doctor with them.

When I first met her she insisted that our interview be held in her garden. Luckily it was sunny and warm that day. I later convinced her to allow me to tour her home on a follow-up visit. She agreed, but noted that I was one of the only persons she had ever allowed into her home over the past several years. *'It's a complete mess,'* she remarked unapologetically.

This proved to be an accurate assessment in my eyes. I had never seen such a *thing-rich* home. It seemed to be made of clutter and nothing more. Yet, possibly thanks to her shooting sticks, Arleen had never fallen. In the course of the visit we toured her upstairs bedroom. There was no exception to the richness of things. Tea bags and a hot-water cooker rested beside an artist's easel, complete with paints and drawings stacked on a make shift table. These competed with other assorted papers, magazines and books piled around the room. The unmade bed exposed a pair of wool gloves, a handheld neck massager and *'ubiquitous cough drops'* that ornamented the linens. These were accompanied by other nearby collections of medication bottles, emery boards, tissues and a pair of nail clippers. However, my view of these objects as clutter shifted as she began to explain their utilities.

Previously, she had called the room her project room. Prior to that it had been the guest room. The room was furnished with a wood-burning stove left uninstalled due to a foreseeable rate increase in homeowner's insurance. She described how the room has once been open and immaculate but over time her projects took over. She was proud of the project table that she had built herself; essentially a door supported by two sawhorses. She went on to describe how her projects grew more and more ridiculous and less and less productive as time marched on.

> Eventually I covered them all up with that sheet to try and reduce the dust on them but it's still a bit weird.' She continued, 'The reason for the stepladder is that… the overhead light bulb burned out and seeing that there are no built-in bookshelves in this house… it now serves both as my light bulb changing spot and my bookshelves. The two beds are matching beds… Once upon a time they were very ornate and orderly with pretty covers and so forth. This was the guest room then but the bed that I was sleeping on in the other room more or less gave out so I moved in here. There is nothing on the bed except the things I need when I sleep like a flashlight and my massager and the ubiquitous cough drops, and oh the gloves in case it's cold. I also make tea up here, especially in the wintertime. I have these two small bed tables which are made for breakfast-in-bed but I use them as writing tables. I put my three three-cornered pillows behind me and put this table on top of me. Then I put the writing machine on that other table and make tea. It's my tea-writing ceremony. Like I said, that's mostly in wintertime. But every now and then, when I just can't stand whatever the challenges are downstairs, I make myself some tea up here. I think I even once made myself some coffee up here.

There are at least two points that emerge here. First is the notion that *one person's clutter is another's treasure* is not entirely accurate. Arleen easily recognized her house as 'a complete mess' and that her veiled projects were 'a bit weird' but she remained proud about the histories and utilities of her cluttered things. This was also evident with Patrick who simultaneously recognized his clutter as 'old, old stuff' but also cherished heirlooms making it difficult to sort and remove. Even more relevant for the argument here is how clutter moves self-care by *nesting* things within arm's reach. Hence there is an ordered efficiency in collecting clutter that economizes body movement. Why go downstairs to make coffee or tea when it can be done upstairs on the spot? At the same time, distributing clutter moves are also apparent. For instance, her partially veiled art and craft projects helped to mix emotional moves of creative identity with memories in the past and present.

The favourite armchair offers another variation of nested collecting-distributing clutter moves. Many of my participants had a favourite place to sit. Often this was a reclining chair or armchair positioned in front of the television. But other views were also monitored. One participant explained how she positioned her chair so that she could simultaneously look at television, eat dinner and watch the neighbour boys play in the yard across the street. Another participant had aimed her chair so she could watch the harbour with its panorama of boats, mountains and sunsets. Yet, besides sites for monitoring the outside world, they were also collection points for home clutter.

Meet Phyllis, in her early nineties. She had just had a stroke a few months prior and was learning to cope with its effects. Her Cadillac was parked in the driveway but she was resigned to the idea that she would never drive again. She was also becoming accustomed to her wheeled-walker. She practised walking without it but was unable to get very far. It framed her body, almost squeezed it, as we sat and chatted. She had positioned it directly in front of her ready for service should she decide to get up. Within arm's length she had placed a side table. It was stacked with numerous objects and I asked her to comment on these. She began by explaining the different publications to which she subscribed.

> I get the daily paper. It comes seven days a week and delivered to the door. My caregiver picks it up and brings it in when she comes in the morning. I also get several magazines but am beginning to get tired of those. I get the Crisis [the official magazine from the National Association for the Advancement of Colored People (N.A.A.C.P.)]. I also get Ebony [a monthly magazine marketed for African Americans]. Oh and I get the Decision from Billy Graham. A lot of this stuff has to do with church. It's my central headquarters. I keep my Sunday school lesson here too. I taught Sunday school at my church before I became like this. I won't say an invalid. I used to teach a bible class.

She explained how she spent a lot of time sitting and praying. She noted a bible that competed for space with a TV remote control, a telephone, a small radio, a fingernail file, a calendar, a check book, a few pens and pencils, a notebook, a box of tissues, a pair of eyeglasses and several other knickknacks all illuminated by a reading lamp. Phyllis continued,

> 'If I hear something on the radio or if something comes to my mind that I want to discuss with somebody or if I want to call somebody I write it down in my calendar.' She started then to look for it. 'I can hardly find it now there's so much here. I'm running out of space.' I then asked, 'So when you're not up and about or in bed it sounds like you're sitting here most of the time.' She replied,
> 'I am here most of the time. Like I said it's my central headquarters.'

While clutter moves helped support Phyllis' failing body in her domestic surroundings, they also helped to facilitate the ordering of her thoughts but also their intersections into the world. Hence Phyllis' armchair headquarters is another illustration of collecting-distributing clutter moves that interweave home ecologies. This was facilitated by assembling together multiple technologies such as her walker, armchair, newspapers, pens, calendar, radio, television and so on. Here the boundary between things and *technologies* in the standard view is dissolved. In the next section the analytical shift from clutter things as *technology clutter* to *clutter technology* is explored in further detail.

From Technology Clutter to Clutter Technology

This section outlines the analytical move from *technology clutter* in the standard or conventional view to *clutter technology* in the socio-technical or symmetrical view. More specifically, *clutter technology* is coined to extend the heuristic of *clutter moves* for rethinking new technologies in relation to old age homecare ecologies. This move stems partly from the paradox mentioned above about how home clutter may be viewed simultaneously as both treasure and clutter or even mess and order. In other words, apparently disordered objects are revealed as meaningful and ordered socio-technical things that support old age homecare.

Technology Clutter

First, let us return to another room of Arleen's clutter. This time we are amidst a collection of various healthcare machines. She begins by surveying the top of her waist-high dresser.

> 'Oh yes I use this area here for combing my hair and storing my hearing aids and putting them in and taking them out and cleaning my glasses.' Then I asked, 'Are these your hearing aids here?' as I point to two black cases. 'Those are the cases for the hearing aids. Would you like to see a hearing aid?' she asked. 'Well okay, you could show me one... Is it going to be hard to put it back in?' I ask. 'No. It's just going to be time consuming because everything is time consuming at my age,' she laughs. 'Also you have to keep track of the batteries and when they die it starts beeping at you. When it does beep this is how you disconnect it so that you're not using the battery when you're not using the hearing aid. There's a button here that you can... That's the wrong button. Hum, no, this button tests that it's on. You can use it when answering the phone or change the frequency. This other button is a simple volume button: You can make it louder, louder, louder, louder! Alright?' she confirmed. 'Okay, thank you,' I replied. 'Well you're welcome. Now you're going to have to wait another day while I put it back in the case. Also I didn't show you the part that sometimes comes apart. There are always a few little things... Like part of it dropping out when you're not looking. Alright what else do you need to know here? Oh the wobbly table. That's the wobbly table,' she remarked. It appeared like she started to lose her balance. 'Careful, careful, hold on,' I urged. I stepped slightly backward and nearly tripped myself over the cord to her hairdryer. 'Oops, do not trip on that. I don't usually have it there but that's my hair dryer. I was using that this

morning,' she explained and then pointed across the room. 'You can see where it says blood pressure monitor?' 'Yes,' I responded. 'Actually if you want to walk over there I'll show you that.' We stepped lightly across the room picking our way around a double bed, stacks of paper and a drying rack with a few clothes still left on its bars. We arrived to a small table and chair in the corner. 'When you do this... you're supposed to have this part of your arm at heart level. I sit here and rest my elbow on the box that this thing came in and then I run it. Want to see how it runs?' she asked. She seemed to be enjoying this. 'Yes,' I replied. She continued again. 'I come in and test my blood sugar and I keep a little record right here. It just went on. Then when you push start it goes up and up and up and you stop it when it's high enough. Then you wait and it comes down and then it starts blinking and it tells you what your reading is.'

Before we left the room she noted several other items. For instance, there were two identical mobile blood sugar monitors for her diabetes, one of which had recently malfunctioned causing her distress on a family dinner outing. Then there was a large rechargeable flashlight that she used when the electricity fails. However the battery was also failing and she had been unable to locate a new one. Next was the base to a rechargeable cordless phone minus the handset. She had misplaced the handset some days before but could not find it with the search button because its battery was also dead. These things intermingled with her collected works of Jung, the psychologist. However, many volumes were missing or scattered elsewhere throughout the house. Finally, there was her old and trusted manual typewriter. It was a memento of her writing sabbaticals up in a Canadian mountain cabin when she was younger. 'No electricity. I had to go in by seaplane,' she explained. However, several of its keys were failing and the ribbon was broken. No matter because she had adopted the desktop PC, well actually three. The first two computers had developed viruses so she had been given a third by a friend. Still she kept the other two as backups. She joked as we left the room, 'Between me and my equipment and my old truck we're all pacing each other to see who and what gives out how much and when!'

Arleen's arrangements challenge the standard view of well-polished medical technologies that deliver unfailing systems for assisted living. Instead, such devices inevitably fail, at least temporarily, or break apart only to merge together home ecologies of old age care. Meanwhile, amidst such mechanical deteriorations, Arleen illustrates how her healthcare devices are thoroughly cluttered to achieve a level of old age home (self) care. With the exception of a misplaced hairdryer cord, this is in stark contrast to the portrayals of home clutter as purely hazardous that predominate in the geriatric-related literature reviewed above. Her story exposes how cluttered devices may actually have specific socio-technical trajectories not readily visible when viewed merely as *technology clutter*, i.e. distinct mechanical devices scattered around the house in meaningless ways. For instance, Arleen's pride and self-reliance was evidenced when she explained the ingenious use of the monitor's box as an armrest while testing blood pressure. The same may be said of

Phyllis' armchair and Patrick's shoeboxes. Each apparent cluttered set of devices have the potential for extended clutter moves, from *motion* to *emotion*. This urges the socio-technical notion of *clutter technology* explored further below.

Clutter Technology

The previous examples have already hinted at how home clutter may serve as a relational nexus for various kinds of *clutter moves* entangling home ecologies of old age care. Phyllis' cluttered armchair headquarters potentially enabled the movement of prayers and thoughts for others while easing her own bodily movements. Patrick's shoeboxes collected sentiments and heritage for prospective moves in his past, present and future family connections. Meanwhile, Arleen's various clutter collections helped ease the *motion* of her failing body as well as the *emotion* of her self-reliance and pride. As such, these cluttered assemblages suggest a kind of *messy interface* or *clutter technology* for moving around socio-technical home ecologies of old age care. Here the term *clutter technology* is offered to denote how cluttered things have the propensity for both permeation as well as the creation of old age homecare ecologies.

In one sense, clutter technology is an analytical move that resembles the suggestion by Bell et al. to 'defamiliarize' the home as a familiar or normalized space; in other words 'making by making strange' (Bell et al. 2005). In turn, they propose that this 'opens up' the home as a conceptual space for design thinking. This is also similar to Dourish's call to 'see like an interface', inspired by Scott's book *Seeing like the State* (Scott 1998), to better conceptualize how humans and technologies are thoroughly entwined in mundane ways (Dourish 2007). The recent *ontological turn* offers additional inspiration. One exemplar is the book *Thinking Through Things* (Henare et al. 2007). Here the authors argue that material things should be explored on their own terms, as heuristics, rather than generic objects or artefacts used to signify some predetermined theoretical concern.

This positioning points an ontological finger at the studies of risk assessment outlined earlier. As argued, these studies have generally left the category of home clutter unexplored or boxed-up as a hazardous risk. Alternatively the *clutter moves* heuristic is employed to reopen and rethink home clutter on its own terms. Similarly, the notion of *clutter technology* has emerged as a means to reconsider technology in its broadest ecological and socio-technical sense.

It is important to note here a handful of recent ethnographic studies in the field of human-computer interaction (HCI) concerned with the design of domestic computer technologies and focused on home clutter (cf. Kirk and Sellen 2008; Swan et al. 2007; Swan et al. 2008). Curiously, these studies are sponsored by a major multinational hi-tech company and they have supported product developments such as *surface* interface devices and *tabletop computing*. These authors tend to focus on the literal collection of home clutter and stress how it is intrinsically linked with people's ideas about home and order. They assess clutter as an essential component of the home or a 'home within the home' (Swan et al. 2008, see also Makovicky

2007). Similar themes related with the link between home and clutter are of course proposed in this chapter, in particular the concepts of *nesting* and *collecting*. To their credit, Kirk and Sellen (2008) have identified the possibility of clutter as *sentimental artifacts*. However, the wider conceptual prospects illuminated by *clutter moves* in old age homecare ecologies are generally overlooked in the HCI literature. Hence, in relation to the standard view of new technology development, the notion of clutter technology proposed here aims to engender further innovative reflections on the future design of assistive technologies for old age homecare.

Final Remarks

In closing, this chapter has employed the heuristic *clutter moves* to embark on the analysis of how movements thread together multiple entities in home ecologies of old age care. Most specifically these have included older people and their cluttered technologies viewed broadly as entangled socio-technical assemblages. After a brief review of gerontology and geriatric-related studies on falling, it was established that such literature generally leaves the category of home clutter unturned. A central tendency in such literature is to figure clutter as a hazard in relation to moving around in domestic surroundings. This was then complemented with a series of ethnographic stories from fieldwork on old age homecare. The clutter moves heuristic helped identify a series of movements, most prominently *collecting-distributing* home clutter. Here the category of home clutter was revealed as part of an intersecting heterogeneous home ecology of things, people and old age care practices that entangle family ties as well as nested selves. The remaining sections developed a conceptual shift from the standard view of *technological clutter* to a more symmetrical view of *clutter technology*. The chapter ends with suggesting that *clutter technology* could be employed to rethink the design of future technologies and their entangled relationships with old age homecare ecologies.

This edited volume proposes to interrogate the relationships between new technologies and emerging spaces of care. The question is then what qualifies as *new*, and in relation to what? The standard view of new technologies generally evokes impressions of invisible wireless digital systems or polished electronic devices: faster, stronger, lighter, cheaper. Modernity at its finest. Apparently there is no end in sight to the many feature variations and embodiments pumped out by the high-tech multinationals every year. Meanwhile, these products appear increasingly similar. *New* does not necessarily seem as new anymore. Arguably, any degree of *newness* is largely a matter of perspective and framing.

It should be evident that this chapter has not adopted the standard view of new technology. Instead the aim has been to position the heuristic of *clutter moves* for the rethinking of old age homecare ecologies. Through this analysis the notion of *clutter technology* has emerged to suggest other ways that materialities are entangled in heterogeneous old age homecare ecologies. If there is anything *new* here then it must reside in the *cluttering* or *messing* (Law 2004) with the notion of new technologies. In other words *clutter technology* is offered to contrast such framings. This attempts to draw attention to the partial failings of material devices while acknowledging their inevitable entanglements with socio-technical ecologies of old age homecare.

On the other hand, perhaps the proposal of clutter technology is not so far off the *new technologies* map. After all, it overlaps with researchers working at the cutting edge of exploratory investigations into cluttered and mundane domestic surroundings. Simultaneously, home-based self-care is one space where socio-technical practices are being transformed. The home has now emerged as a prominent site for care as welfare institutions around the world shift healthcare responsibilities to the individual. Subsequently the home is increasingly equipped with healthcare devices and transformed into care facilities in their own right. The socio-technical notion of *clutter technology* aims to offer avenues for the further exploration of such transformations. For instance, in line with the suggestion from Bell et al. (Bell et al. 2005), perhaps clutter technology begins to enable the *defamiliarization* and further expand the conceptual space of old age homecare ecologies. Here the analytical attention on how clutter moves figures prominently.

Bibliography

Bell, G., Blythe, M. and Sengers, P. 2005. Making by making strange: defamiliarization and the design of domestic technologies. *ACM Transactions on Computer-Human Interaction*, 12(2), 149–173.

Blythe, M. A., Monk, A. F. and Doughty, K. 2005. Socially dependable design: the challenge of ageing populations for HCI. *Interacting with Computers*, 17(6), 672–689.

Börsch-Supan, A. et al. 2005. *Health, Ageing and Retirement in Europe: First Results from the Survey of Health, Ageing and Retirement in Europe*. Mannheim: Mannheim Research Institute for the Economics of Aging (MEA).

CDC. 2007. *The State of Aging and Health in America, Centers for Disease Control and Prevention* (CDC).

Collins English Dictionary. Sixth edition. 2003. Glasgow: HarperCollins.

Douglas, M. 1966. *Purity and Danger.* London: Routledge.

Douglas, M. and Wildavsky, A. 1983. *Risk and Culture*. Berkeley: University of California Press.

Dudley, D. 2007. Conquering clutter. *AARP The Magazine*. Jan. and Feb.

Emmett, R. 2005. *The Clutter-Busting Handbook*. Canada: Doubleday.

Evans, G. and Kantrowitz, E. 2002. Socioeconomic status and health: the potential role of environmental risk exposure. *Annual Review of Public Health*, 23(1), 303–331.

Ewer, C. T. 2006. *Houseworks*. London: Dorling Kindersley.

Gershon, R. et al. 2008. Household-related hazardous conditions with implications for patient safety in the home health care sector. *Journal of Patient Safety*, 4(4), 227.

Haraway, D. 1997. M*odest-Witness@Second-Millennium.FemaleMan-Meets-OncoMouse*. New York: Routledge.

Henare, A., Holbraad, M. and Wastell, S. (eds). 2007. *Thinking Through Things*. London: Routledge.

Jansen, S. and Löfving, S. (eds). 2009. *Struggles for Home*. New York: Berghahn.

Kaufman, M. 2006. *Organize It!* New York: Filipacchi.

Kirk, D. and Sellen, A. 2008. On human remains: excavating the home archive. *Microsoft Technical Report*. MSR-TR-2008-8.

Latour, B. 1987. *Science in Action*. Cambridge: Harvard University Press.

Latour, B. 1999. *Pandora's Hope*. Cambridge: Harvard University Press.

Law, J. 2004. *After Method*. London: Routledge.

Makovicky, N. 2007. Closet and cabinet: clutter as cosmology. *Home Cultures*, 4(3), 287–309.

Marshall, S. et al. 2005. Prevalence of selected risk and protective factors for falls in the home. *American Journal of Preventive Medicine*, 28(1), 95–101.

Messecar, D. et al. 2002. Home environmental modification strategies used by caregivers of elders. *Research in Nursing and Health*, 25(5), 357–370.

Mol, A. and Law, J. 1994. Regions, networks and fluids: anaemia and social topology. *Social Studies of Science*, 24(4), 641–671.

Moylan, K. and Binder, E. 2007. Falls in older adults: risk assessment, management and prevention. *The American Journal of Medicine*, 120(6), 493.

Oxford English Dictionary. Second edition. 1992. Oxford: Oxford University Press.

Pfaffenberger, B. 1992. Social anthropology of technology. *Annual Review of Anthropology*, 21(1), 491–516.

Russell, M. et al. 2006. Falls risk and functional decline in older fallers discharged directly from emergency departments. *Journals of Gerontology Series A: Biological and Medical Sciences*, 61(10), 1090.

Schillmeier, M. and Heinlein, M. 2009. Moving homes: from house to nursing home and the (un-)canniness of being at home. *Space and Culture*, 12(2), 218–231.

Scott, J. 1998. *Seeing Like a State*. New Haven: Yale University Press.

Swan, L. et al. 2007. Containing family clutter, in *Home Informatics and Telematics: ICT for the Next Billion, 241, IFIP International Federation for Information Processing*, edited by Venkatesh, A. et al. Boston: Springer, 171–184.

Swan, L., Taylor, A. S. and Harper, R. 2008. Making place for clutter and other ideas of home. *ACM Transactions on Computer-Human Interaction*, 15(2), 1–24.

Tsing, A. L. 2005. *Friction*. Princeton: Princeton University Press.

Urry, J., 2007. *Mobilities*. Cambridge: Polity.

Wherton, J. and Monk, A. 2008. Technological opportunities for supporting people with dementia who are living at home. *International Journal of Human-Computer Studies*, 66(8), 571–586.

Wilson, P. and Rodgers, B., 2006. Research on falls prevention and physical activity in older adults and a notice of a new web-based quality system by the agency for healthcare research and quality. *Home Healthcare Nurse*, 24(10), 632–636.

Zecevic, A. et al., 2006. Defining a fall and reasons for falling: comparisons among the views of seniors, health care providers, and the research literature. *The Gerontologist,* 46(3), 367.

Zecevic, A. et al., 2009. Utilization of the seniors falls investigation methodology to identify system-wide causes of falls in community-dwelling seniors. *The Gerontologist*, 49(5), 685.

Chapter 5

Homespace or Workspace?
The Use of Multiple Assistive Technologies in Private Dwellings

Hanne Lindegaard and Søsser Brodersen

Introduction

Sophie is living with her husband in a house they bought twenty years ago, in a suburb north of Copenhagen. Ten years after they bought the house, Sophie was diagnosed with Multiple Sclerosis. After living with the disease for nine years, Sophie and her husband realized last year that she was no longer able to manage the stairs leading to the first floor, and they decided to add a new bedroom and bathroom to the ground floor construction. Due to her disabilities, the family was entitled to receive assistance and advice from the municipality's architect. When Sophie and her husband saw the architect's proposal for rebuilding their ground floor, they were shocked. They had expected solutions that would meet Sophie needs and wish to be more self-sufficient in her home, but the proposal was based on her assumed future needs for intensive assistant care. The proposal described a house with ceilings prepared for a track hoist system for lifts, a bedroom prepared for a hospital bed, and a bathroom three times larger than normal size. Thus, the house was transformed from a home for Sophie and her family to an assumed workspace for future care providers. The house would become a 'hybrid space' – a home, but also a dwelling that would accommodate being packed with assistive technologies designed to assist Sophie as well as the care providers.

This chapter is based on the ethnographic research project, 'Homespace or Workspace?', and analyzes the socio-material relations in the use and distribution of assistive technology in Denmark. The material was gathered from 2007 to 2009. The research project used field studies and interviews with multiple actors from the Danish Sclerosis Society, the municipality, domiciliary care-people, the family etc. During the research period, we visited ten households, applying a mixture of qualitative methods such as observations and qualitative interviews. The observations were made in the homes where we 'followed the actors' in their daily duties and practices (Bijker 1995, Shove et al. 2008). Here, we observed *how* and *when* the disabled, their families and the domiciliary care-people used or did not use the assistive technologies, and *whom* the technologies actually assisted.

In the research and the analysis, we have also used historical documents, documents containing information about health care and design practices as well as material about a new Danish concept in social welfare, 'Welfare Technology'. This concept covers a range of new technological solutions (remote-controlled door/window openers, toilets with douche and drying, robots to assist eating etc.) to be tested for people with disabilities. The main strategy is to find out if this kind of technology enhances the independence of the severely physically and cognitively disabled in their homes (e.g. Jordansen 2009).

The study analyzes how a socio-material space for humans and non-humans – here the private home – changes when it becomes a hybrid of workspace and private space (e.g. Latour 1999). How does everyday practice change, both in the interplay with the institutionalized artefacts but also with the multiple actors involved in domiciliary care? How are the assistive technologies shaped in the interplay with everyday life in private dwellings? In Denmark, disabled and elderly people are able to receive domiciliary care and assistive technologies from the municipality. But how do different users integrate, refuse or negotiate such assistance and artefacts? How do the 'human' and the 'non-human actors' enter the homes? Do the artefacts stay as 'guests' or will they be tamed and domesticated as part of everyday living? Will they become a part of an aligned network and even homey?

The chapter is also concerned with an ongoing discussion of care practices and its relation to both 'being and feeling at home' and the ongoing standardization of the practice of care (e.g. Schillmeier, M. and Heinlein, M. 2009; Fisher 1991). Schillmeier et al.'s article 'Moving Homes' presents a German male, 'Mr. B', who had two strokes. After the second stroke, he was forced to move from his private flat into a nursing home. The study shows that there is much heterogeneity in care practices, which involve different kinds of assistive technologies. This influences *how* and *when* 'Mr. B' feels 'at home' in the nursing home, where changing professional health care routines, bodily positions and more technologies influence his feeling of becoming more disabled and of 'uncanniness'. In our study, we examine the private home as a space where continual optimization of professional health care practice involving many assistive technologies often collides with the disables' expectations concerning the kind of non-human actors they prefer in their homes. It is obvious that number, functionality, style and scripts influence whether a dwelling feels homey or rather like an institution.

The Theoretical Framework

The theoretical framework and the terminology in the analysis are drawn from Science and Technology Studies (STS), where Actor Network Theory (ANT) is the key inspiration. Here, the complexity of socio-material relations and the concept of 'scripts' are in focus (e.g. Latour 1999; Akrich 1992; and Woolgar 1990). Domestication Theory (e.g. Silverstone 1989; Lie/Sorensen 1996) is also used.

Actor Network Theory (ANT) was developed by Bruno Latour, Michel Callon and John Law in the 1980s. It is characteristic of ANT that it does not distinguish between human and non-human actors. In ANT, the human and non-human actors interact in an interwoven network. The term 'hybrid' is used to address the heterogeneous relations between all the human and non-human actors involved, e.g. in the dwelling. In the notion of the human/non-human hybrid, the combination of actors and the assistive technology is in focus (e.g. Latour 1999).

According to Latour (1999), the relation between actors and artefacts is not stable; therefore, the chapter explicitly discusses the dynamic relation between complexes of material artefacts, conventions and competences with focus on everyday life. It is not only the relation between specific artefacts and individual users, as analyzed in a usability study; analysis is also made of how collections of artefacts co-evolve, and also how the different actors in the actor network have different expectations concerning how, where and whom the artefacts will assist. In this study, it has been very useful to explore how and when the actors and the various artefacts were able to become an aligned network or *when* and *why* this failed.

The domestication theory draws attention to the aspects of meaning of artefacts, and how the symbolic relates to practice. This means that a domestication analysis focuses on use, but also goes beyond function and use. What is happening after the artefacts leave the designers' desks or the municipality service? Do the actors e.g. integrate the assistive technologies provided by the municipality in their homes and everyday life, and how does this process actually occur? Is it a linear progress without conflicts or negotiations? If not, what kinds of negotiations can be identified? Domestication covers both the processes in which technology is adapted to everyday life, and the processes that involve everyday life's adaptation to the technology. The analysis emphasizes micro relations in everyday settings, and focuses on how the general symbolism of the artefacts' codes can be converted into something personal, connected to identity and social relations (Lie/Sørensen 1996).

In a domestication analysis, the process is specified in up to four phases: appropriation, objectification, incorporation and conversion. The appropriation phase occurs when the artefact – i.e. the assistive technology – is provided for the physically and mentally disabled. Then, through objectification, the assistive technology is given its place and made visible. In the third phase, the artefact is incorporated into the daily routines; and in the fourth phase, conversion, the user also manifests the strong artefact-actor relation involving identity and values to the outside world. A domestication analysis examines the cultural integration – or disintegration – of artefacts, and sees it as a process in which technology, actors and space are affected.

Sophie's House: Workspace or Home Space?

When we first visited Sophie, she met us outside her house sitting in her electric wheelchair. The front yard, door and entrance did not differ from other houses, and nothing indicated that this was a wheelchair user's home. Sophie gave us a guided tour of her house, starting on the ground floor in the kitchen, dining room, living room, and utility room. She then told us to go upstairs to the first floor without her, because she was not able to do that herself any longer. While we were upstairs, Sophie waited for us downstairs, guiding us from that distance. On the first floor, we saw two bedrooms and a bathroom. Until last year, Sophie had been sleeping in one of the bedrooms together with her husband. Now, since her Multiple Sclerosis had intensified, she was unable to climb the stairs, and had therefore added a new bedroom and bathroom to the ground floor. When we came downstairs again, Sophie guided us to the newer part of the house. Sophie seemed very happy and proud; she said that the addition/annex now was quite as she liked it. The ceiling had a dormer with a remote controlled window in the sleeping area, and there were sliding doors to a terrace, where Sophie could go out herself. The bathroom was

> …twice the size that I need, but I was not able to convince the municipality consultant about this argument. He told me that in a few years I would be happy that there was lots of room for my *disabled care*… that stung.

After the guided tour, we sat drinking coffee in the dining room, right next to the annex. During the interview, Sophie explained how the planning process of the annex had occurred. She did not hide the fact that she had been very upset about it and sad, when she saw the architect's first proposal for the annex. The ceilings were flat so that later it would be easier to install a lift. The idea was that the annex should be prepared for the future course of Sophie's illness so that she could receive the care and means of assistance that her care demanded;

> …but I couldn't live with those ceilings. Think that during the next maybe ten years I was supposed to lie there and look up and think that I would surely get worse and finally have to be hoisted with a crane from my bedroom to my bathroom – that was impossible.

This argument was listened to, and Sophie's wishes were met.

During our interview and our visit at Sophie house, it became clear that Sophie refers to herself as 'wheelchair user'. When she uses her wheelchair in her daily routines in her home, it is a routine use in which the relationship between Sophie and her wheelchair has developed into a strongly aligned network (Latour 1999). Sophie has incorporated the technological object. As a spokesperson for the Danish Sclerosis Society, she expresses this aligned network to everybody, and her identity and status are as a 'wheelchair user'. Sophie and her wheelchair act as 'hybrids'.

Sophie's relation to the new bathroom and bedroom plans were quite different. It was not an aligned network; the network is much broader, with a multiple set of relations. For Sophie, the new bathroom symbolizes her possible future need for help, and it is designed for the municipality network with the care people as its main users. The plans for the house were designed for her as a disabled person whose future needs demanded assistant care. They were not designed for Sophie as a self-sufficient woman. She looked forward for the new bathroom, where she saw herself and her wheelchair as the main users, and where it was all right that the basin was height adjustable and that there was room for a bathing chair to assist her.

But the super-size bathroom with the in-scripted mechanical lifting system was built as a workspace from the perspective of the architect and the municipality. The objective was to allow room for lifts in order to prevent occupational injuries for professional nursing assistants in the future. It was designed for multiple-use practices involving many human and non-human actors. This became a dilemma that is a major challenge for architects and designers – to make the dwelling part workspace, part home. For Sophie, an institutionalized setting was proposed where the disabled and her family could hardly feel at home.

This is a huge dilemma. Homecare aims to make it possible for people to remain at home rather than to live in institutional settings. Both the assistive technology and the staff are supposed to help people to bath, dress, move etc. To do the job, the care providers often need assistive technologies (e.g. patient lifts) in their daily work practice to avoid work-related injuries. In Denmark, staffs do not bring assistive technology with them; therefore, assistive technologies must be stored in the patients' homes. This was the reason Sophie's home was supposed to be rebuilt in such a way as to provide room for all the future assistive technologies and for the assistants to maneuver with lifts, hospital beds and toilet chairs.

During the interviews with Sophie, it became clear that she sees herself as a self-sufficient woman and not as a patient who needs lots of help from the care sector. Sophie can do most of her daily routines herself, with the help of her electric wheelchair, which she really appreciates. For Sophie, the wheelchair is a non-human artefact that is domesticated for most of her everyday routines. It assists her in her daily routines and provides her with self-help, so she can maneuver inside and outside the house in her everyday practice. This kind of assistive technology meets the UN Standard Rules on the Equalization of Opportunities for Persons with Disabilities, which state:

States should ensure the development and supply of support services, including assistive devices for persons with disabilities, to assist them to increase their level of independence in their daily living and to exercise their rights.[1]

In Denmark, the Danish Centre for Assistive Technology develops and improves the quality of assistive technology by helping to ensure that standardized technological solutions are designed, developed and applied in a way that benefits all users:

---

1 United Nations, 2006: Standard Rules on the Equalizations of Opportunities for Persons with Disabilities. www.un.org/esa/socdev/enable/dissre03.htm

> *Assistive technology should do more than assist. It must be a natural part of everyday life and meet users' needs and preferences both aesthetically and functionally.*[2]

But even though the Danish Center for Assistive Technology aims to help people with disabilities to be more independent in their daily lives, observations from our study indicate that many of the assistive technologies are made to help the staff in their care practices and optimize their workspace. Since 1975, working environment rules were consolidated into the Danish Working Environment Act, which applies to all work performed for an employer. The central aspect of this legislation is the extended safety and health concept, which means that all factors causing accidents, sickness and attrition must be taken into consideration in prevention work. The legislation covers such areas as work performance, workplace design, technical equipment etc.:

> *The workplace of each individual person shall be expediently designed and fitted out and be so spacious that all necessary furniture, equipment and materials can be placed in a safe way in relation to each other and in such a way that all functions in connection with the performance of the work can be carried out safely and with safe work postures and movements.*[3]

Floor space, room height and room capacity shall be adapted to the nature of the work, the technical equipment, materials and furniture in the working area, as well as the number of persons normally entering the area. The legislation applies to all employers, also the people working in private dwellings, also to Sophie's bedroom and bathroom plans.

The Patient Lift

Many assistive technologies are thus not designed to assist the disabled but to meet the requirements for health care practice and assist the nurses. The mobile patient lift provides an example; it is obvious that it is neither designed for the patient nor for use in private homes.

If we look at the history of lifts, we find that the first floor lift for moving patients was patented in 1955 with the title, '*Floor Crane with Adjustable Legs*'.[4] The design was based on a similar device used in automotive repair shops to lift '*engines and other heavy parts*'.[5]

---

2   www.hmi.dk.

3   The Danish Working Environment Legislation, www.at.dk/sw12173.asp.

4   US Patent 2706120.

5   US Patent 2706120.

In hospitals and nursing homes, this type of lift has been domesticated as the way to move all patients and thus reduce work-related injuries. The staff is no longer allowed to move patients without using lifts.

Today, even modern floor lifts maintain the basic elements of the original 1950 design described in the patent, and engine lift practice is still in-scripted in the patient lift. According to Akrich, technical objects '...*define a framework of action together with the actors and the space in which they are supposed to act*' (Akrich 1992). This socio-material space is therefore very important, both when choosing assistive technology and when designing it.

The huge difference between lifting an engine and lifting a patient emerges when examining and analyzing the socio-material network of which the lifts are a part. The context and practice are not quite equal. The engine lift moves an object (a car engine), while the patient lift (assisted by a nurse) moves a subject (a disabled person) – an important and obvious difference that can be understood by analyzing the socio-material practice.

The practice for using the patient lift is that the patient hangs in a sling made of textile fabric while being lifted between the bed and the wheelchair, the wheelchair and the toilet, or while turning the patient in bed. To use the lift, the patient must lie on a couch or in bed so that the care provider can place a net underneath the patient's back. Metal rings are fixed to each of the four corners of the rectangular net. To lift the patient, the crane's arm raises the corners of the net to lift the patient. The patient thus hangs in the net without being able to help the lifting process, which often places the patient in uncomfortable positions.

The mobile lift is designed to be 'user friendly' – but only for a specific group of users, not all users. Many patients feel very uncomfortable when hanging in the patient lift; interviews with patients and assistants indicate that the script of the lift means that the patients are made to be passive objects. The script of the lift does not allow patients to be anything else but 'patients', i.e. passively waiting for help.

Schillmeier et al. compare the mobile patient lift with a sit-to-stand-lift, which Mr. B's nurses used. Mr. B sits at the edge of his bed, and the lift is placed in front of him, a short distance away. A belt is lashed to the upper part of Mr. B's body, while his feet are placed in a basin-like anti-slip feature. A nurse pushes a button and the lift automatically begins to pull Mr. B slowly into a standing position. Mr. B has to help by pressing both his feet firmly into the anti-slip feature. After some weeks, Mr. B. can only manage to stand on one leg, and the nurses are then forced to use the mobile lift. In this lift, Mr. B has nothing left to do.

Although many patients are lift 'users' every day, the lift is not domesticated as an incorporated artefact, and the relation between the artefact and the patient is not an aligned network. In interviews with domiciliary care people, they often mention that some disabled persons cannot understand what the technology can do for them, and they meet the assistive technologies with skepticism.

For the nurses, the lift has become a domesticated artefact that they use in their daily routines in the workspace, and for the domiciliary care people from the municipality, it has become an artefact they distribute to disabled patients at home.

Patient lifts are therefore stored in private dwellings, in bedrooms, bathrooms, or even in living rooms, waiting for the care people to come to use them. In our research, we found a new design, the LikoLight, which is a portable mobile lift '*developed for people in need of a light, mobile lift that fold away easily*'. The LikoLight works much like any other mobile lift, but it is portable and can be conveniently stored; it even has a carrying bag.[6] With this lift, it is possible for care providers to bring the lift with them, instead of storing it in the home.

The Dilemma of Institutionalized Design in a Homey Space

It is obvious that in the process of re-designing the engine lift, the patient has not been considered the main user, and the patient's 'comfort' and 'independence' have not been considered design criteria. The designers have not inscribed the complex socio-material context into the lift, nor is it either aesthetically or functionally designed to be used in a private home.

In Denmark, home care practice for the disabled is part of the welfare policy, administrated by local municipality. In 2007, Denmark spent three billion Danish kroner (40 million Euros) on assistive technology, a figure that is expected to increase in coming years because of the aging population.

> This implies a growing need for technological solutions to enable people with disabilities to be active participants in society and maximize their quality of life.[7]

Resistance to Assistive Technology

Alice, who was diagnosed with Parkinson's, lives in a house with her two children. Alice is one of about five per cent of Parkinson's patients who are diagnosed before the age of 40. Alice is very self-sufficient and has a minimum of assistive technology in her home. She has a bathing chair, a trolley, crutches, and a wheelchair to use when she goes to the shopping center with her children. The day we visited Alice, she had just received a tele-care device that enabled her to call for assistants at night. Due to her illness, Alice has muscular rigidity, and she sometimes has trouble turning over and changing positions in bed. Before she received the device, she used a Bariatric Trapeze, a triangular device hanging above her bed:

> ...but the Bariatric Trapeze was scaring away my new boy friend. He peeled it off and I have not mounted it again.

---

6   www.progressivemobility.com/products.asp.
7   www.hmi.dk.

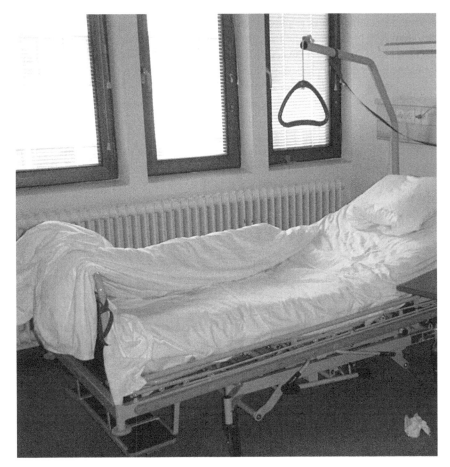

Figure 5.1   The Hospital Bed

Alice has a double bed, and she told us that the trapeze helped her change positions at night, but the design and script reminded her of a hospital or nursing home. Her children did not like it either. After the trapeze was removed, Alice explored the assistive technology market to find an alternative:

> The occupational therapist told me to ask the municipality for a hospital bed with guard rail ... but I am not that disabled. I love my bed and do not want to sleep alone for the rest of my life...

For Alice, it is not a quality of life to have a Bariatric Trapeze or a hospital bed in her home. It reminds her of her handicap and visualizes it, and frightens her family and boyfriend. The trapeze is originally designed for hospitals and does not fit into traditional homes – either in size or style. The artefact is designed for an institutional context, and requires that Alice both has to reorganize her home and her identity. The script of the trapeze and the hospital bed is so domesticated into the hospital context that it is very difficult to bring it into a private sphere. Although it has many functional properties, the script is not open. Hospital practice inscribes that hospital beds are made for hospitalized patients and are compatible with non-human actors like bed tables, hygiene standards, long corridors, longevity etc. A hospital bed is not designed to be part of a private home, and for Alice and her family, it did not work.

Care people from the municipality have a standard list of possible artefacts to suggest; most of them are not trained to analyze the broader practices. They try nicely to find the best solution for the disabled, but they often only consider the isolated functional properties and not the whole socio-material network.

Concluding remarks

Our research emphasizes that especially 'quality of life' is very important for the disabled. For Sophie, her electric wheelchair, her adjustable basin, and her bathing chair really maximize her quality of life, whereas some of the other assistive technologies in her home had not become domesticated. Artefacts that are to assist the care providers could, for instance, be 'hidden' when not in use, or could be designed so that it was easier for assistants to bring them to the home themselves. Assistive technology design thus embeds particular expectations concerning purpose, context, practice and use. The scripting of the assistive technology is most obvious when artefact designs configure the user in specific and practical ways. The concept of scripting highlights the range of contextual, practical, material and semiotic elements that need to be taken into account when analyzing actual use (Shove et al. 2007). Although an artefact is prescribed, the scripts remain open when exposed to the hybrid use practice. Artefacts undoubtedly script and configure use practices, but the user also reconfigures the artefact – objects and users configure each other.

But how is it that nobody has designed a comfortable double bed for disabled people living in private homes with their families? Why it is difficult to re-think and re-design new assistive technology to 'assist' all the human and non-human actors in the socio-material network? Design for disabilities in home space could be a major future design possibility. The hybrid space between humans and non-humans opens the way for new lines of enquiry regarding the role of artefacts in health care. The dwelling is a hybrid entity, but for some disabled people, their dependence on many different assistive technologies makes their houses resemble an aggregate of machinery. Both the quantitative aspect (the numbers) of the technologies and the style of the artefacts influence what it means to live in a well-functioning home space. Designing assistive technologies demands that developers, designers, care providers and politicians acknowledge the complexity within which assistive technologies are to be used and applied. This requires understanding who is to use which technology and where. Solutions and systems need much more consideration, so that they take into account all the different users' practices and wishes.

References

Akrich, M. 1992. The Description of Technical Objects, in *Shaping Technology/ Building Society*, edited by in *Shaping Technology/Building Society*, edited by W. E. Bijker and J. Law. Cambridge: MIT Press.

Bijker, W. 1995. *Of Bicycles, Bakelites and Bulbs – Toward a Theory of Sociotechnical Change*. Cambridge, Mass.: MIT.

Fisher, B. J. 1991. *It's not quite like home: Illness career descent and the stigma of living at a multilevel care retirement facility*. New York. Garland.

Gray, D. B. et al. 1998. *Designing and Using Assistive Technology*. Brookes.

Jordansen, I. K. 2009. Velfærdsteknologi – selvhjulpen med ny teknologi. Hjælpemiddelsinstituttet.

Latour, B. 1999. *Pandora's Hope*. Cambridge: Harvard University Press.

Lie, M. and Sørensen, K. (eds). 1996. *Making Technology Our Own? Domesticating Technology into Everyday Life*. Oslo: Scandinavian University Press.

Oudshoorn, N. and Pinch, T. 2003. *How Users Matter. The Co-construction of Users and Technologies*. Cambridge: MIT.

Schillmeier, M. and Heinlein, M. 2009. Moving Homes: From House to Nursing Home and the (Un)Canniness and Being at Home. *Space and Culture*, 12(2), 218–231.

Shove, E., Watson, M., Hand, M. and Ingram, J. 2007. *The Design of Everyday Life*. Oxford and New York.

Silverstone et al. 1989. Families, technologies, and consumption: The household and information and communication technologies. CRICT discussions paper. Brunel University.

Woolgar, S. 1991. Configuring the User: The Case of Usability Trials, in *A Sociology of Monsters: Essays on Power, Technology and Domination*, edited by J. Law. London: Routledge.

Chapter 6

Electric Snakes and Mechanical Ladders? Social Presence, Domestic Spaces, and Human-Robot Interactions

Mark Paterson

Introduction: The Robot in the Home

Since the first industrial robotic arm became operational in 1961 at General Motors, the International Federation of Robotics (IFR) reports that the global robot population has grown to 6.5 million. Of these, 5.5 million are unmanned vehicles and service robots (Guizzo 2008). The rate of robot population increase is stunning, and is expected to reach 18.2 million by the year 2011. Turning to human populations, the forecasted demographic shift will be especially pronounced in Japan, Europe and the United States. This demographic shift occurs alongside the increasing number of operational service robots, robots that directly aid human activities as opposed to industrial robots in factories. In other words, human interactions with robots are becoming increasingly ubiquitous, and robots are leaving the military and factories and are being invited into peoples' homes. While partly driven by a profit-motivated robotics industry inevitably seeking diversification, the governments of these postindustrial, high-tech, high-density nations with their proportionately aging populations and coincidentally high levels of consumer spending, their service sector economies and the emphasis on knowledge transfer from higher education research institutions, all point to an exponentially increasing population of service robots that spill over from heavy industry and the military, and start to invade the home *en masse*. This is happening already most noticeably in the toy market, ever since 'must-have' Christmas toy crazes like the Furby™ and My Real Baby™, and now with an array of robot toys and interactive pets that range from horrendously complex and expensive, such as Lego Mindstorms or Sony's AIBO, down to simple, cheap and cheerful such as Wowwee's Roboboa. Robots are everywhere, and there is a burgeoning ecology of "machines to live with" (Brooks 2002; Thrift 2004: 470). Furthermore, as robots are diversifying in their nature, reach and abilities, so the potential client groups are similarly diversifying. For years, industrial workers have been dealing with robots in spatial proximity, but other sectors of society such as the mobility impaired and the elderly will benefit from assisted living through robotics, involving physical assistance, rehabilitation and other therapeutic uses of robotics.

The question here is an important one. Just what kind of social presence do robots currently have, and what kind do we *want* them to have? To answer this, we look at robotics and human interaction through a series of case studies, and structure the chapter according to themes pertinent to social presence of robots in assisted living contexts. Firstly we look at a grounded, mundane example of an available technology that, like most robotics, originates in a military-industrial context but which is currently receiving much interest from the UK government for future domestic use. This is a remotely-controlled 'snake-arm' robot designed for use within confined spaces. Secondly, from this existing example and its context we look to the purpose and policy of service robots in general, examining competing strategies and approaches in Europe, the US and Japan. We will see that the same potential demographic problem is being approached in different ways, ways that reveal more of the cultural-historical construction and reception of robotics in these geographically disparate locales. Thirdly, the varying and contentious role of 'presence' and 'proximity' of such service robots will be explored in the settings envisaged for them. Contrary to popular imagination and much science fiction, few robots are currently certified as 'safe' to be working alongside humans, and ISO (International Organisation for Standardisation) safety certifications for robots have become a hot topic in Europe and the US recently (Bicchi et al. 2008). The mixed and sometimes confined spaces of physical human-robot interactions (pHRI) is therefore timely and important to consider. Throughout the treatment of these examples, themes and sociocultural analysis, one concept persists throughout: space. For the mixed spatialities of national policies and responses to demographics, the cultural and safety coded responses and expectations of confined spaces of human-robot interactions, and the mixed use of robotics within industry, the military and now increasingly the home, all must address the shifting, sanctioned and increasingly proximal spaces of human-robot interactions (HRI). The final section looks at the 'nature' of these spaces of interaction. For, in order that non-traditional client groups willingly accept physical assistance from robots, the unfamiliarly machinic and non-humanoid cannot appear as uncanny. Within these mixed human-machine spaces, movements and gestures must somehow be naturalised, the *unheimlich* become homely, the machinic appear naturalistic.

The Case of 'Snake-Arm Robots'

Prior to discussing the shifting social contexts of an alien technology in the home it would be worthwhile to look at a more down-to-earth example of a robotics solution, and note the confluence of management decisions, government policies and research funding directives that are prizing a robot away from the military and industrial applications it was originally designed for, and into the domesticated spaces of the home. A global survey conducted by the United Nations and International Federation for Robotics usefully divides robot populations into three categories: industrial robotics, professional service robotics, and personal service robotics (UN and IFFR 2002). These three categories describe different technologies, application domains and contexts. While industrial robots were the first production robots, the service-based categories are more recent. A professional service robot exists outside an industrial context, assisting people in the pursuit of their professional goals such as a surgical robot working to enhance a surgeon's skills in an operating theatre, or the HelpMate® robot performing the mundane physical task of distributing food and medication in hospitals. Personal service robots, on the other hand, "assist or entertain people in domestic settings or in recreational activities", explains Thrun (2004: 12). This is the biggest growth area, with autonomous lawnmowers and vacuum cleaners. Famously, Rodney Brooks' company iRobot makes the Roomba, an autonomous self-guided robot vacuum cleaner. The snake-arm robot had its origins as an industrial robot, but its planned transition into a personal service robot, with all the social complexity that entails, is the focus for this chapter.

But what is a snake-arm robot anyway? The majority of the ever-burgeoning population of robots discussed earlier are technically 'robot arms'. In industrial settings, robot arms are programmed with repetitive movement and orientation tasks, and provided spatial coordinates within which to operate through computerised routines and subroutines. Robot arms, like human arms, have points of articulation that allow up to six degrees of freedom (DOF) and hence can operate within a bounded three-dimensional space. The space is usually limited by the fixed base that the arm is attached to, a base that provides power and servo control. Snake-arm robots are different. They can be mounted on a moving base with wheels or tracks, and so are mobile rather than fixed like robot arms. More significantly, there are so many articulations that it is more akin to a snake skeleton or human spine than a human arm:

> [A] snake-arm robot is a bit like the human spine. It comprises a large number of vertebrae. It is a tendon driven arm with wires terminating at various points along the length of the arm. The result is that the curvature and plane of curvature of each segment can be independently controlled. A motor is used to control the length of each wire independently. The control software calculates the necessary lengths of all the wires to produce the desired shape. (Buckingham and Graham 2005: 127)

The increased articulations allow many more degrees of freedom (DOF) within a fixed space than the conventional robot arm. Furthermore, because each segment is controlled independently, the key difference from the conventional robot arm is the ability to plot a more sinuous point-to-point plotted path through a given three-dimensional space. This is why snake-arm robots are also known as 'continuum robots', being a continuously curving manipulator, much like an elephant trunk in being freely prehensile. Or, reporting what the developer of the robot Rob Buckingham clarifies, "This robot has no 'elbows,' which allows it to 'follow its nose' while maneuvering in tight quarters", unlike conventional industrial robots, which are virtually useless 'because their elbows get in the way'" (Buckingham, in Rutherford 2008). This difference is crucial in severely confined spaces, and therefore these snake-arm robots are particularly suited to applications and settings where it is dangerous, difficult or simply impossible for humans alone to access. So far, these have been military-industrial applications including access to dangerously radioactive and confined nuclear reactor cores and chambers, bomb disposal and removal duties, and aircraft assembly in the confined spaces of the fuselage.

The development of snake-arm robots is fairly recent and somewhat of a commercial and engineering niche. Research on snake movement and robotics by Howie Choset at Carnegie-Mellon University for example has explored some biomimetic aspects of locomotion and path planning, and are more correctly described as 'serpentine' robots rather than snake-arm robots. Related projects at Carnegie-Mellon have included prehensile devices more akin to elephant trunks. But the history of snake-arm robots (as opposed to snake-like or serpentine robots) is to some extent coterminous with the development of the UK-based company OCRobotics, developer and manufacturer of the first production snake-arm robots. The company came to prominence in this field and caught the eye of the engineering and science press when, faced with a major engineering problem in the nuclear power station Ringhals 1 in 2003, the Swedish government needed a way to access a chamber underneath the reactor core and repair a pipe, or face shutting down the plant completely. The chamber was impossible to access by human engineers alone, as along with the exposure to radiation the space was further confined by wide tubes that contained the control rod drive mechanisms (CRDM). After conducting extensive simulations the snake-arm robot could access these spaces through remote control by a human operator from a safe distance, yet with a 50 micron [millionths of a metre] degree of accuracy (Buckingham and Graham 2005: 124) and, due to the individually controllable reticulated spines through their own servo motors, creating 23 degrees of freedom. The repair being successful, the previously unproven snake-arm robot gained interest from, amongst others, the Department for Trade and Industry (DTI, now the BERR) in the UK and the US Department of Defense. In the words of Buckingham, the snake-arm robots had "proved themselves able to 'reach the unreachable'" (Buckingham and Graham 2005: 127).

While the significance of the snake-arm robot is clear for industry and the military there are far wider implications for the population at large. 'Reaching the unreachable' in hazardous industrial contexts or potentially dangerous military situations means the ability to work within confined spaces, and therefore potentially hybrid spaces of human-robot proximity. While currently the snake-arm uses remote control through a camera feed at the end of the arm, the technology for more autonomous movement, for sensing obstacles and altering movement through a robotic 'skin', is currently being developed (Fildes 2006). Both pre-programmed and autonomous snake-arm movement can be utilised by a wider population requiring physical assistance in the more confined domestic spaces of the home or care institutions. More detailed domestic scenarios are envisaged in a later section.

A number of factors are converging here: demographics, the economic-industrial imperative to transplant military technologies to other milieux, and the large grants that government agencies are paying for this to occur, effected through knowledge transfer schemes (in the UK for example the Technology Strategy Board, the DTI). Recently the news has been full of 'swarm' robots (e.g. Palmer 2008), with competitions involving school and university students to build small autonomous robot systems that interact and distribute tasks amongst the swarm. The swarm leads to robust technologies that achieve their tasks even if individual robots get damaged or destroyed, and the emphasis is therefore on keeping the devices simple, reducing the cost of manufacture, making them robust. Or in the words of robotics pioneer Rodney Brooks, and the title of a paper and a documentary film in which he and his robots featured, "fast, cheap and out of control". The 'out of control' aspect allows emergent patterns of behaviour to occur. It will be a while before emergent behaviours and movements will be allowed in domestic settings, perhaps. But once the preserve of large industrial corporations or the military, the 'fast' and 'cheap' aspects of robotics are eminently applicable to the proliferation of electronic toys and gadget robotics currently being marketed, becoming noticeably more sophisticated in nature and complexity in the past decade or so. This argument is explored by Thrift (2008), so I will not rehearse it here. Instead, we concentrate on another aspect of the proliferation of fast and cheap robotics into the marketplace, one pertinent to the case of snake-arm robots: so-called 'service robots'.

Service Robots: Machines to Live With

Rodney Brooks envisions the penetration of robots into the home along the same lines as the now ubiquitous presence of computers (2002: 113). Likewise, Bill Gates has said that personal robotics today is at the stage that personal computers were in the mid-1970s. Having personally overseen and influenced the goal of a PC in every home with Microsoft co-founder Paul Allen, in an article on personal robotics for *Scientific American* Gates claims the next step is 'a robot in every home' (Gates 2007). Having likened the current state of robotics to that of the PC industry in the 1970s, Microsoft co-founder Bill Gates assumes that particular applications of robotics will drive them into mass acceptance and finally into the home. Although Gates cannot predict what applications these will be exactly, he suggests:

> It seems quite likely, however, that robots will play an important role in providing physical assistance and even companionship for the elderly. Robotic devices will probably help people with disabilities get around and extend the strength and endurance of soldiers, construction workers and medical professionals. Robots will maintain dangerous industrial machines, handle hazardous materials and monitor remote oil pipelines. They will enable health care workers to diagnose and treat patients who may be thousands of miles away, and they will be a central feature of security systems and search-and-rescue operations. (Gates 2007: 64)

Indeed, he goes on to argue that robots will become so ubiquitous and specialised and affordable for consumers that they will be unrecognisable as the anthropomorphic robots of science fiction. Gates uses the term 'devices', but a more fitting word for robots within domestic contexts might be 'appliance'.

In thinking about the robot as appliance, effectively personal service robots designed to do low-level tasks of physical assistance, there are requirements for human-robot interaction due to the shared and confined nature of those spaces. What Gates presumably had in mind are personal service robots for recreation or for personal assistance, rather than professional service robots. Yet healthcare institutions and care homes will certainly benefit from both personal service robots and professional service robots alike; personal service robots for physical assistance, and for emotional presence and amelioration of the effects of Alzheimers, like Paro the robot seal (Wada and Shibata 2008; Broekens et al. 2009) for example. But there is also the need for professional service robots that, as appliances, look unlike any human or animal. Robots like HelpMate®, 'Pearl' the nursebot (Pineau et al. 2003), or the recently announced Panasonic drug delivery robot, one of a new breed of medical robots that has the ability to sort medical supplies and deliver them to trained staff in hospitals, but not administer them to patients yet. Resolutely non-humanoid and as yet nameless, according to a Panasonic spokesman theirs "looks like a cabinet with lots of small drawers" (ABC News 2009). Previously, the Humanoid Robotics division of Japan's government-backed National Institute of Advanced Industrial Science and Technology earlier had revealed HRP-4C, a walking, talking robot with a female face. In news stories it was presented as a catwalk 'fashion model' robot and even appeared on the catwalk at Tokyo's Fashion Week, but like its predecessor HRP-2 was originally designed as a less glamorous domestic helper robot. "But it hasn't cleared safety standards and cannot yet help humans with daily chores", reports Associated Press (ABC News 2009). As we have seen, this is only a matter of time.

Some evidence suggests that a humanoid form or appearance is more reassuring (Kiesler and Hinds 2004), and this effect must be more pronounced in unfamiliar or clinical surroundings such as healthcare settings. However, naturalistic interaction can occur through establishing interaction and movement protocols whatever the robotic form, as we shall see. Bringing together several of the factors mentioned above, including the burgeoning ecology of 'machines to live with', the diversification of client groups due to demographic shifts, such as the increased population of the elderly and the relative paucity of immigrants to care for them, and the phenomenon of more complex human-robot interactions (HRI) within industrial and increasingly domestic settings, along with the reduction in the cost of manufacture, there is an inevitability to the increased number of relatively low-cost robots in the home. Such robots are 'service robots', and are symptomatic of a familiar shift from technologies developed at great expense for the military which now find a market within consumer electronics.

Apart from toys, genuinely useful robots are already being purchased for the home. The most famous and successful example is iRobot Corporation's self-guided vacuum cleaning robot for the home, Roomba, which returns itself to a charging station when necessary, managing maximum coverage of the floorspace not through clever programming, GPS, distributed waystations or homing signals. Instead, it plots a random course and detects the proximity of walls, covering the floorspace through sheer reiteration and minor variation. In other words, like any appliance it is not 'clever' but performs a domestic task effectively and robustly. iRobot continues to make robots for the US military but is diversifying its domestic line, complementing the Roomba, first produced in 2002, with the Scooba floor-washing robot in 2006 and the Verro pool-cleaning robot in 2007. These are the first service robots that are being accepted into domestic spaces precisely because of their ability to act as *appliances*, pursuing physical activity that reduces the need for mindless and repetitive human manual labour. Given this rapid development and uptake of service robots, the demographic shifts specified previously, and the proven nature of domestic consumer market and rapid increase in robot populations through toys and appliances such as the Roomba, we are now in a position to ask about the exact nature of the spaces of human-robot interaction. The public appetite for such low-end toy robots and relatively uncomplicated robotic appliances has been demonstrated through healthy sales figures, with three million Roombas sold worldwide as of 2009 (iRobot.com 2009). But what of more complex and physically larger robotic appliances designed not for amusement or light work duties but for everyday physical assistance? It might not be productive to extrapolate data from the sales of simple appliances, and the clients targeted will be different. But as regards the spaces of human-robot interaction, we ask: how are non-traditional client groups to negotiate with robots' everyday spatial presence? How will this occur in domestic spaces, and what is the nature of the mixed or hybrid spaces of human-robot interaction that result? To answer this we follow two concepts in particular through from their deployment in human-computer interaction (HCI) to human-robot interaction (HRI): presence and proximity.

Hybrid Spaces of Human-Robot Interaction (HRI)

Before discussing the specifics of the newly-emerging area of HRI, some comparison with the field of Human-Computer Interaction (HCI) is worthwhile. HCI has a venerable history, arguably starting from Vannevar Bush's visionary 1945 paper 'As We May Think', but an explosion of research from the 1980s onwards shows a great deal of crossover with the social sciences, including the use of ethnography and Garfinkel's ethnomethodology in computer-supported collaborative work (CSCW), usability studies and interface design (see e.g. Card et al. 1983 for an early introduction to the psychology of HCI; Dix et al. 2003, the 3rd Edition of their textbook; Myers 1998; Dourish 2004 for overviews). Briefly, HCI is a broad field encompassing interaction design, ergonomics, software and hardware engineering, and is increasingly moving beyond the keyboard and mouse input devices pioneered during the 1970s at Stanford Research Labs and Xerox

Palo Alto Research Center to incorporate areas such as gesture recognition and augmented reality (e.g. Myers 1998: 52). Conversely, HRI studies are only slowly becoming established, as Kiesler and Hinds (2004) show. Mutual influences abound between these two areas of interaction, and commonalities include the breadth of the fields, since HRI itself is influenced by human-computer interaction but also artificial intelligence, robotics engineering, natural language understanding and the social sciences. Both share an emphasis on usability, that is, the ease with which an operator can accomplish a set task or, in HCI terms, how the user can 'articulate' an 'operation' within a system (e.g. Dix et al. 2003: 128ff). But some concerns are specific to HRI. The basic goal of HRI is to produce principles and algorithms allowing more natural and effective interaction between humans and robots, in proximity or through remote, long distance teleoperation of unmanned vehicles. Following methodological cues from HCI, researchers in HRI take future interaction design cues and protocols from observations of current patterns of interaction between humans and machines, and this branches into specific areas within HRI such as 'motion planning', computing trajectories within spaces of human-robot interaction that avoid collision. This is made more complex by the dynamic state of the space of HRI, when robots and/or humans are mobile, and also through the relatively restricted number of degrees of freedom (DOF) that humanoid and non-humanoid robots currently exhibit. Therefore a fundamental question is beginning to be asked within HRI: with the rapid increase in robot populations, how are non-traditional robot-using client groups to negotiate with robots' everyday spatial presence? And, given the traditional industrial contexts of HRI, how will this occur in more domestic and everyday spaces? In the case of OCRobotics, large grants from DTI to transplant military-industrial robotic technologies into other areas, including the home.

The seeming inevitability of robot domestication should not blind us to important counterarguments, and my position is not straightforwardly as apologist or uncritical celebrant of this process. However, firstly, for purposes of physical assistance and rehabilitation it is clear there are non-trivial applications for this technology. In the face of concerns over health and safety, workloads and a shortage of nursing staff in hospitals and carehomes, novel sources of physical assistance are welcome. And secondly, more generally, it stresses the importance of human-robot interaction (HRI) within the next decades. A large proportion of the working day in industrialised countries is spent with computers and screens where the human-computer interface (HCI) is being continually refined and rethought for usability purposes. Likewise, near-future scenarios of dwelling with or alongside service robots and other machines for rehabilitation inevitably begs the question of *how* we interact with them, and what kind of human-robot space of interaction results. For this, we can consider two areas, the character of the hybrid spaces of human-robot interaction, especially between non-traditional client groups, and the character of the 'presence' – and especially *social* presence – of these service robots.

Robot Presence: 'Presence' and 'Social Presence' Through Interaction

Through technologies of human-computer interaction and human-robot interaction alike, the requirement even over a long-distance network is to grasp, move, perform direct manipulation with an immediate real-time effect, and with clearly perceivable feedback. Low latency and fluidity in response, and even anticipation of the user's movement, helps foster the perception of interacting with something real and tangible (see e.g. Paterson 2006; 2007). In other words, engineering a sense of 'presence' can be said to occur through both the human-computer interface (HCI) and the human-robot interface (HRI). These unfolding technologies are a set of augmentations that begin to play with an emerging multisensory realm where presence is 'felt', one that talks of the engendering and engineering of 'immersion,' of 'presence,' of 'aura' through the addition of touch or an engagement on multiple sensory levels. The human-computer interface (HCI) has explored ideas of presence on the near-space of the computer desktop, using graphical displays and, more recently, haptic technologies. As Dreyfus argues, "what gives our sense of being in direct touch with reality is that we bring about changes in the world and get perceptual feedback concerning what we have done" (2000: 57), perceptual feedback that can equally be produced through visual and haptic technologies on the near-space of the computer desktop, or distant spaces through teleoperation. Whether near space or distant space, a virtual object has a sense of presence through real-time interaction, what HCI literature terms 'direct manipulation' (e.g. Schniederman 1987). But the term 'presence' has a particular history within computer-mediated communication and later virtual reality, deriving from research into remote operation and 'telepresence'. In 1980 Marvin Minsky at MIT defined telepresence in terms of the manipulation of objects in the real world through remote access technology (Steuer 1993). Shortening the term simply to 'presence,' Sheridan (1992) applies this not only to controlling real-world objects remotely, but also as a descriptor for the felt effect of people interacting with and immersing themselves in virtual reality or virtual environments. The advent of the internet's global networked reach, increasing numbers of computers, and the promise of greater interactivity collectively detached the notion of presence from specialized teleoperation contexts, and presence became applicable to everyday computer-mediated communications.

The sense of presence that results from bidirectional visual, haptic, oral or gestural communications occurring through distance is 'telepresence'. Whereas currently telepresence has been a selling point for low latency audio-visual videoconferencing applications, used for example in the marketing of Tandberg and Cisco Systems, its origins invoke a more haptic language of direct manipulation of objects, of getting a grip, bringing the hands as well as the eyes into play. Evidently, the very notion of 'telepresence' is premised on assumptions of what 'presence' itself is. Or, as Dreyfus so usefully reminds us: "For there to be a sense of presence in telepresence one would have to be involved in getting a grip on something at a distance" (2000: 58). In one of the original and clearest formulations of the need to think about 'presence' and 'telepresence', early VR commentator Michael Heim considered robotic telepresence:

> Virtual reality shades into telepresence when you are present from a distant location – 'present' in the sense that you are aware of what's going on, effective, and able to accomplish tasks by observing, reaching, grabbing, and moving objects with your own hands as though they were close up. [...] Robotic telepresence brings real-time human effectiveness to a real-world location without there being a human in the flesh at that location (1993: 114).

Both military and medical applications for robotic telepresence technology quickly became apparent, the ability to conduct surgery over protracted distances, whether a remote battlefield or simply a civilian medical outpost far from specialised facilities. In this case, telepresence becomes the ability to accomplish tasks, perform operations or manipulations at a distance with a convincing enough visual and/or haptic perception of interaction with real-world objects.

Away from the interactive space of networked computer presence, robot interactions occur in hybrid human-machine spaces, depending on the human perception of a non-human (machinic) presence. Robotic presence, as above, may still imply tangible engagement with a human operator or dweller, either through remote operation displayed on a screen, or in the mixed space of human-robot interaction through observations and responses to movements and gestures. Engineering and social science literature alike concur that facial movements in particular elicit affective response (e.g. Breazeal 2003b; MacDormand and Ishiguro 2006), which is examined further in the following section. But generally, allied with Milgram's reality-virtuality continuum (in Holz et al. 2009), we might equally posit a continuum of robot presence, starting with 'appliances' or unsophisticated robots like Roomba™ that perform routine and repetitive tasks without significant autonomy or self guidance. Stepping up the level of robot presence, non-humanoid service and healthcare robots that interact and dwell routinely within human spaces such as the HelpMate®, must exhibit sets of predictable movements and interactions, especially if they appear more humanoid, to be deemed 'acceptable' and non-threatening by non-traditional users. Another step up in robot presence applies to humanoid robots exhibiting more highly articulated human-like responses or movements, such as Korean Institute of Science and Technology (KIST) project Mahru, or the facial expressions in University of California San Diego's Einstein head, with thirty servo motors corresponding to thirty facial muscles (see Fox 2009). As we will see, perceiving even a minor mismatch between action and appropriate response touches upon the psychology of the uncanny, not only in terms of the aesthetics of humanoid appearance but also in terms of the flow of movement and response that answers a human action or command.

If presence and telepresence have traditionally been considered as crucial concepts for VR and human-computer interaction, teleoperation and computer-mediated communication over networks then, the possibilities for HRI are in the early stages. Another factor pertinent to human-robot interaction is the engineering of *social presence* (see e.g. Lee 2004; Lee et al. 2006; Spexhard et al. 2007), the extent to which robots come across as social actors within human-robot interactions. Robots can achieve this through responding seemingly appropriately to various sensory stimuli, and responding with corresponding gestural actions such as meeting one's gaze or responding to a voice command. Robots such as Asimo and BARTHOC exhibit this behaviour, and MIT's Cog will shortly be examined in this respect. In social presence, robots as 'virtual actors' are experienced as 'actual social actors' in either sensory or non-sensory ways (Lee 2004). Understandably, most attention is paid to humanoid robots in this regard. If the vast majority of the current robot population consist of disembodied robot arms employed in industrial settings, social presence need not be an issue. However, robots with humanoid form, like Honda's ASIMO or Sony's QRIO, or recognisable humanoid body segments such as the upper torso (e.g. the recent European ECCEROBOT), or simply the production of recognisably human facial expressions, (e.g. the Repliee-Q2 and BARTHOC), are projects whose goals explicitly instantiate the biomimicry of a range of human actions and communications to enhance social presence within human-robot interactions.

In short, a major reason for concentrating on recognisably humanoid facial articulations and limb and motion patterns, despite the complexity of the engineering, is to ease these machines into more acceptable proximal interactions with humans in a range of spatial contexts, including healthcare and assisted living. While children and science fiction fans are typical early adopters of novel technological equipment, the more pressing needs of physical assistance and labour requirements in mixed environments, especially when it comes to the elderly in so-called 'assisted living' projects, suggest a positive correlation between recognisably humanoid robot form and immediate social presence. Especially in Korea and Japan, the preponderance of humanoid robots signals exactly this. Yet other social experiments with non-humanoid virtual actors involving biomimicry reveal alternative forms of interaction. Paro the therapeutic robot seal was the subject of study at a Japanese care home, reportedly encouraging affective encounters between residents and the playful seal (see also Wada et al., in Kidd et al. 2006), although it is unclear if an American or British context would elicit similar responses. Such "electric animals" (Thrift 2008) may attend to multiple human needs, including mechanical assistance for those with age-related physical impairments, can function as mechanical equivalents for animal-assisted activities (AAA) or animal-assisted therapy (AAT), and through a series of interactions including expressive movements in face, eyes and limbs may even fulfil emotional requirements. Given the increasing popularity of robot animal toys like Sony's AIBO at the younger end of the spectrum, and healthcare robots at the older end, it seems inevitable that robots will become increasingly present in the home, and therefore that both humanoid and non-humanoid robots must become 'social' or even 'sociable' in order for their presence to seem more naturalistic and reassuring to non-traditional client groups. What might these machinic forms of sociality or sociability entail?

Sociable Robots: Human-Robot Interactions

After her interactions with a selection of robots including current projects within the Humanoid Robotics Group and the Personal Robotics Group both at MIT, the home of Cog, Kismet and more recently Mertz, in a long journalistic piece for the *New York Times* Robin Henig intentionally describes these as 'sociable robots'. In observing and designing human-robot interactions some researchers consider their robots as 'social', but another researcher at MIT, Breazeal, prefers the term 'sociable' (2003a). Henig acknowledges that a proportion of these interactions are unnaturalistic, break the illusion of human-like interaction, and due to hardware breakdown or software glitch the carefully-honed illusion of naturalism can easily break back down into uncanniness. There is something sad and pitiful in a broken-down robot, and such robots are "still less like thinking, autonomous creatures than they are like fancy puppets that frequently break down". Further explaining what makes such robots 'sociable' rather than 'social', she concludes:

> Sociable robots come equipped with the very abilities that humans have evolved to ease our interactions with one another: eye contact, gaze direction, turn-taking, shared attention. They are programmed to learn the way humans learn, by starting with a core of basic drives and abilities and adding to them as their physical and social experiences accrue. People respond to the robots' social cues almost without thinking, and as a result the robots give the impression of being somehow, improbably, alive. (Henig 2007: 31)

In other words, the illusion of 'aliveness' (an appearance fostered through movement, which underlines properties of an animate spirit within clockwork machines or automata, for example) is an emergent property derived from a diverse set of naturalistic-seeming human-robot interactions, with a large degree of biomimicry. Setting aside any parallels between developmental psychology and robot learning, there is a qualitative difference between a 'social' robot that allows a communicative pathway between human and robot through speech or gesture recognition, for example, and a 'sociable' robot that appears to look at and listen to you, and performs a dynamic series of seemingly appropriate gestural or vocal responses.

Brooks, head of the Robotics Lab at MIT was co-inventor of the now infamous robots Cog and Kismet, robots programmed with a series of low-level subroutines to respond to certain stimuli, yet which exhibited fairly complex emergent patterns of behaviour nonetheless. Their vision and auditory subsystems were basic yet together these robots appeared 'alive'. The Cog robot was basically an upper torso with camera eyes and arms. In 1995 Sherry Turkle was researching how children consider a machine as 'alive' in the face of increasing numbers of toys and machines that "speak, respond to stimuli and seem to have an ongoing inner life" (Brooks 2002:149). When Turkle visited Brooks' Robotics Lab and interacted with Cog, she surprised herself at her reaction to its 'aliveness':

Cog 'noticed' me soon after I entered its room. Its head turned to follow me and I was embarrassed to note that this made me happy. I found myself competing with another visitor for its attention. At one point, I felt sure that Cog's eyes had 'caught' my own. My visit left me shaken – not by anything that Cog was able to accomplish but by my own reaction to 'him'.

Turkle then goes on to admit: "I had behaved as though in the presence of another being" (Turkle 1995: 266). The question of anthropomorphism, and why a more naturalistic reaction might occur with humanoid as opposed to non-humanoid robots, points to some potential difficulties for the acceptance of a snake-arm or serpentine robot. Margaret Boden (2006) considers the role of a convincing-enough animism, given that the technological level required to excite anthropomorphic responses, as Turkle discovered personally above with Cog, is not that great. If such a sliding scale remains crude or inexact there is certainly a correlation between early clockwork machines, toys and automata that moved and mimicked human or animal motions, such as Vaucanson's mechanical Digesting Duck of 1738, up to a more anthropomorphic set of complex behaviours and interactions, another example from the world of automata being the Mechanical Turk unveiled in 1770 by Wolfgang von Kempelen, but later found to be a hoax (see e.g. Wood 2003).

Having discussed presence and social presence in human-robot interactions, how might this relate specifically to our earlier example of snake-arm robots? Can we talk effectively of an artificial ethology? If we think of non-traditional user groups, such as within assisted living environments, how can these mechanical actors produce a 'presence' that is decidedly non-threatening and non-aggressive? Can this be achieved more effectively through more accurate modelling of animal ethology? Can such biomimetic behaviours extend effectively to non-humanoid robots? Can a sense of animal-like presence be literally engineered through the appropriate patterns of movement, constituted by arrangements of sensors and actuators? In many ways, this seems the converse of the lifeless automaton. Wood's history of automata, couched as a quest for 'mechanical life' (2003), plays largely on the notion of mechanical life as mimicry of 'natural' behaviour by artificial mechanisms. As such, the Freudian notion of the uncanny becomes prescient once again.

The 'Uncanny Valley'

Ernst Jentsch's concept of 'the uncanny', identified in his 1906 essay, 'On the Psychology of the Uncanny' (1995), was famously taken up and elaborated upon by Freud in his 1919 essay 'The Uncanny' (1985). It should come as no surprise that for both Jentsch and Freud, a sensation of uneasiness or unpleasantness accompanies a mechanical toy, doll or automaton that is *too* lifelike, especially if on a human scale. Jentsch explains: "A doll which closes and opens its eyes by itself, or a small automatic toy, will cause no notable sensation [of uncanniness], while on the other hand, for example, the life-size automata that perform complicated tasks, blow trumpets, dance, and so forth, very easily give one a feeling of unease." Then, importantly for this argument, Jentsch continues: "The finer the mechanism and the truer to nature the formal reproduction, the more strongly the special effect [i.e. uncanniness] also makes its appearance" (Jentsch 1995: 12). Both Jentsch and Freud explore some parallels between the forms of empathic identification and uncanny alienation between humans and toys or automata, and then apply this to the kinds of empathy and affects of 'lifelike' performance and characterisation in theatre and storytelling. However, the implications of the uncanny in mechanical toys and automata is ripe for exploration in terms of the evolution of robotics. Japanese engineer and roboticist Masahiro Mori foresaw the difficulty of increasing biomimetic verisimilitude with his notion of the 'Uncanny Valley' (Mori 1970).

While contemplating how robots were evolving towards more recognisably humanoid forms and movements, Mori observed a correlation between how 'lifelike' a robot appears and how we identify with and empathise with a robot, until a certain point. For the sake of argument we might empathise with a robot which is 20 per cent 'humanlike', more so with a robot which is 50 per cent, and even more with a robot 90 per cent lifelike. Mori plotted out a rising slope of anthropomorphizing empathy. However there is a precipitous drop-off after around 95 per cent into what he terms the 'Uncanny Valley'. Approaching increasing perfection in biomimetic verisimilitude might incidentally question the ontological status of 'robot' and start to invoke perhaps another category, 'replicant' (as occurs notoriously in Ridley Scott's 1982 film *Blade Runner*). In such cases the slightest variance, the incorrect 5 per cent, looms up disproportionately, rendering the assemblage somehow creepy and monstrously alien. Following the curve further, as the appearance and motion of the robot continue to become less distinguishable from a human being the emotional response becomes positively correlated once again, approaching human-to-human empathy levels, thinks Mori (1970: 34). The precipitous drop and subsequent dramatic rise in empathy and identification with a robot, if plotted on a graph, visually demonstrates this 'valley'. The uncanny valley is the area of repulsive response aroused by a robot with appearance and motion between a 'barely human' and 'fully human' entity. The name captures the idea that a robot which is 'almost human' will seem overly strange or alien to a human, and thus fail to evoke the empathic response required for productive human-robot interaction. Hence, in terms of the design of a proposed robot working in confined spaces of human-robot interaction, the combination of appearance and movement protocols must negotiate around this valley. Given that potential client groups are likely to find such machines alienating, the task is a difficult one, and while it is tempting to suggest that the perception of robot forms is predominantly a concern for the elderly, Woods (2006) has shown that children's perception of robotic form is resolutely aware of uncanny or alienating factors. Along with the facial and limb gestures of humanoid robots, with a verisimilitude that approaches the human form, we might apply this hypothesis not only to non-humanoid robot forms such as snakes or bears like RIBA, but also as Weschler (2006) shows, to realistic 3D computer animation where an increasingly popular tranche of films that blend Computer-Generated Imagery (CGI) and live action must similarly avoid the 'uncanny valley'.

'Maybe the Robots will Take Care of Us': A Scenario for Healthcare Robots

Putting these strands together, we can see the need for service robots as non-complex appliances that have a limited form of non-threatening social presence within an assisted living context. Such assisted living environments will be mixed or hybrid spaces of human-robot interaction. As Kiesler and Hinds point out, one of the most significant and novel uses for robotics will be replacing the low-level duty of increasingly sparse healthcare professionals: "For example, nurses making rounds in assisted living facilities spend much of their time sorting and administering medications. A robotic assistant could do some of this work, as well as chores that are difficult for elderly people such as fetching newspapers and mail, getting up and down stairs [etc …] enabling elderly people to be independent longer" (2004:2). But how exactly might such robots be deployed in this context? In this section we envisage a scenario based on current technology demonstrations of robotics from various labs worldwide. There are plentiful examples of service robots devised for medical assistance and, rather than an exhaustive survey, a selection of representative instances shows the relevance of the concepts discussed so far, and will conceptually provide a realistic scenario for the deployment of snake-arm robots.

The RIBA (Robot for Interactive Body Assistance) was reported in 2009 as a modification of a model developed in 2006 called the RI-MAN (Robot Interacting with Human). Developed by researchers in the Bio-Mimetic Control Research Center of Japanese organisation RIKEN, RIBA was designed primarily to assist nurses by lifting patients in and out of beds and wheelchairs, and on and off the toilet. As one commentator summarises, the RIBA is intended to function as a "robot nurse", and although it weighs 400lb itself "can reportedly lift patients up to 135lbs out of bed or a wheelchair, while also making use of a full range of tactile sensors and some special 'soft skin' material to ensure the short journey is as comfortable as possible" (Melanson 2009). The multi-jointed arms are embedded with tactile sensors that optimise the lifting and carrying of patients and, like human arms, yield somewhat to physical pressure thereby reducing the brute contact between machine arm and human flesh, and ensuring a more naturalistic carriage and posture. RIBA is designed therefore to operate exactly within these mixed spaces of human-robot proximity, and addresses one of the perennial difficulties for trained nursing staff, the heavy lifting of human patients, with its concomitant health and safety concerns. One feature of RIBA's interactivity with humans is a response to voice cues and commands. As another commentator explains, "Using visual and audio data from its surroundings, RIBA can identify co-workers, determine the position of those nearby, and respond flexibly to changes in the immediate environment" (Pink Tentacle 2009). For some inexplicable reason, with its gangly multi-jointed arms, the robot has a non-humanoid disguise, with the head of a teddy bear. Instead of legs the torso tapers down into a platform with omnidirectional wheels (like the iRobot Roomba), enabling movement and rotation in confined spaces, around hospital floors and nursing facilities. While the arms mimic human joint articulation and movement patterns, the result is a melange of non-humanoid, humanoid and robotic forms. In a press release by RIBA's creators RIKEN, the teddy bear appearance was chosen to put patients at ease, and attempting to make the robot humanoid in appearance would only frighten people.

However, it is unclear what its targeted client group, the elderly, will make of it. Seemingly a bizarre or even misguided project, this robot is a small yet significant step in the acceptance and presence of service robots in healthcare settings. RIKEN plans to have these robots in production and operating within hospitals and nursing homes in the next five years. The faultlines for the various socio-economic conditions and policy decisions in place for this to occur are thereby revealed. For in Japan, where the RI-MAN and RIBA have been developed specifically for healthcare purposes, instances of robots developed by major corporations and government sponsored research with significant investment is an attempt to directly address the impending demographic timebomb where an aging population, a decline in young people, and therefore a shift in the ratio of nurses and carers, is being tackled predominantly through technology. This has been a clear policy focus for Japan in recent years. If figures published by the US-based National Center for Policy Analysis (NCPA) are to be believed, the proportion of the population over 65 in Japan will jump from 18 per cent in 2002 to 30 per cent by 2030, and 37 per cent by 2050, a significantly larger jump than the remaining top ten economically advanced nations in the OECD (Hagist and Kotlikof 2006). This has inescapable implications for healthcare spending, where the proportion of GDP spent on healthcare in Japan will rise from 6.7 per cent in 2002 to 18.2 per cent by 2050, a large proportion of this to be spent on technological solutions. The rapidly-changing nature of the healthcare situation in Japan is summarised by Ienewski:

> The advent of self-care and self-management support programs, the boom of developing mobile solutions for personal health monitoring, personal health coaches obtaining immediate feedback, and health information from patients at home are strong indicators today for a silent revolution of an aging society reshaping the health care and well-being landscape (2007: 222).

As the population increases, and the proportion of GDP spent on healthcare increases, there will be a continual mismatch of healthcare provision to the population, a low ratio of healthcare professionals to an aging population, and long-term economic growth will inevitably suffer. The healthcare shortfall is only bridgeable through significant current and future investment in technology. Even today, enterprising private healthcare solutions such as 'Medical Malls' are springing up in major urban centres like Tokyo and Osaka to service cash-rich but time-poor consumers, such as executives and managerial office workers, according to Greenberg (2008). These stopgap solutions arise to a chronically slow governmental response, where there remains no official policy to the rather stark demographic projections. According to Gross (2009), the ratio of working-age to elderly Japanese "fell from 8 to 1 in 1975 to 3.3 to 1 in 2005 and may shrivel to 1.3 to 1 in 2055". Kiyoaki Fujiwara, director of economic policy at the Japan Business Federation, is understandably pessimistic: "In 2055, people will come to work when they have time off from long-term care" (in Gross 2009). Given the absence of official governmental plans for social engineering to reverse the overall decline in population and the shifting ratio of workers to elderly, or healthcare professionals to the general population, it seems the only solution is technological. Japan retains its lead in engineering and technology, two pillars of its phenomenal postwar economic growth, even if social engineering cannot address this problem head on. In other words, the focus, rather fatalistically, is on technology to manage the inevitability of demographic shift rather than on altering it through other means. It is worth remembering these demographic factors, the fatalistic attitude and the lack of a coordinated governmental policy response, as it puts the following vignette into context. On a visit to Toyota's vast and astonishingly automated Tsutsumi plant known as 'Toyota City', with legions of robot arms doing 90 per cent of the welding work on a car production line with a capacity of 400,000 vehicles per year, Gross asked a city official how demographic changes would affect the delivery of health care. The official responded, only partially in jest: "Maybe the robots will take care of us" (Gross 2009).

Given the demographic figures, the lack of coherent healthcare policy, and the rather dystopian vision of Toyota City with its millions of robot arms, the uncertain place of robots in Japanese healthcare is salutary. The EU-funded Ambient Assisted Living (AAL) fund is determined to meet similar demographic and infrastructural challenges that will inevitably hit Europe down the line. One of the projects, a European answer to Japanese healthcare robots like RI-MAN, is the CompanionAble project for elderly care, based in Reading, UK from 2008. Not an all-singing, all-dancing humanoid robot, it consists in a more synergetic solution involving a mobile robotic companion within an associated "ambient assisted living environment" that supports both family and professional carers in their daily tasks, including cognitive stimulation and therapy for the recipient. Thus the rather humble mobile robotic companion works collaboratively with a 'smart' stationary home environment, facilitating human healthcare presence and communication including gerontologists and cognitive therapists, in the case of dementia. The mobile robot aspect acts as mediator and companion but is continually interacting with its smart environment, the space having multiple sensors that monitor the health of inhabitants, essentially plugged into a whole healthcare social network that promotes communication between the care recipient and their professional helpers. Press coverage and technology websites abound with inconsequential stories from East Asia of unicycling robots, dancing robots or robots that mimic human facial expressions through servomotors, and the emphasis tends to stay on individual robots' abilities to replicate movements or gestures that are culturally meaningful to us. But by shifting focus from the single robot figure itself, CompanionAble is revealed as a series of nodes in a more socially-integrated environment. The mobile robot figure is continually in contact with the care recipient and professional human carers, and the larger built environment with its distributed sensors. One stated aim of the project is "To achieve the continuous availability of sense-ful close support and cognitive engagement of the elderly" (Badii 2009). Not simply a sociable robot, this is a sociable environment that actively encourages human interactions, cognitive engagement, and a reassuring 'sense-ful' contact. Still in its early stages, CompanionAble seems a far cry from the depressing vista we often assume, of robots only encouraging social isolation amongst the elderly. These forms of human-robot interaction are not limited to one-to-one encounters, then, but show the promise of plugging the care recipient back in to a caring social milieu.

Conclusion: Robotic Skin, Robotic Touch

For the sake of argument, let's take a hypothetical step forward. Like RIBA, the rather cute bear with the human arms, add a tactile layer of skin. Its arms adapt to your bodyweight, it 'feels' when you are in place and it carries you to the specified destination. If there is too much pressure at one point in the arms, adjustments are made. Now, add a skin to the snake-arm robot so that it becomes a skinned snake-arm robot (but not a snake-skinned robot arm). The 'Snake Skin' project is the next step in the evolution of OC Robotics' snake-arm robot, having won a UK government contract:

> The project is to develop an intelligent 'skin' for snake-arm robots that will not only provide a barrier between the arm and the environment, but will also have the ability to sense the presence of obstacles by touch or proximity. Other capabilities will be explored, including the measurement and control of temperature, the sensing of chemicals, and any environmental changes that are relevant to the end user. (OCRobotics.com)

The sensing, feeling snake-arm robot is an appliance whose time will come. Having painted the demographic picture in East Asia, Western Europe and the US, professional service robots like RIBA, the CompanionAble and a skinned version of the snake-arm robot will simply have to be there for us, as physical assistance. Currently they are complex appliances with a low level of autonomy. However, with the kinds of incremental step that include skinning, and therefore the ability to perceive touch within the confined spaces of human-robot interaction, our domestic spaces will become transformed. We can only envisage the multifarious possibilities for new forms of social presence through programmed 'naturalistic' (i.e. biomimetic) behaviours, movement protocols that respond to particular patterns of responses to human actions and movements. Some interesting questions arise. If the tactile layer senses a human limb in its path, perhaps additionally confirmed through its vision system, what would be a safe way for it to react? What paths of movement can we expect, if their forms are borrowed from animal behaviours? If the arm moves in a serpentine way, would this be expected or simply alarming? For the sake of safety in confined spaces, a dance will have to be worked out, a pattern of movements, of actions and reactions to human bodily position and limb movement, that future service robots will have to work out. As toys and pool-cleaning gadgets our robots are currently at a pet-like scale, but with their growth in size, population and presence in the next decades, our domestic spaces will be transformed. And how we interact with these machines will have to be worked out. This may be as much of a dance between social scientists and software programmers as between snake-arm robots and human limbs.

References

ABC News. 2009. Panasonic breaks into robotics with medical robot. [Online 7 July]. Available at: http://abcnews.go.com/Technology/Health/wireStory?id=8019883 [accessed 15 August 2009]

Baadi, A. 2009. CompanionAble: Integrated Cognitive Assistive and Domotic Companion Robotic Systems for Ability and Security. PDF flyer [Online]. Available at: www.companionable.net [accessed 19 January 2010].

Bicchi, A., Peshkin, M. A. and Colgate J. E. 2008. Safety for Physical Human-Robot Interaction. in *Springer Handbook of Robotics*, edited by B. Siciliano and O. Khatib. New York: Springer, 1335–1348.

Breazeal, C. 2003a. Towards sociable robots. *Robots and Autonomous Systems*, 42(1), 167–175.

Breazeal, C. 2003b. Emotion and sociable humanoid robots. *International Journal of Human-Computer Studies*, 59, 119–155.

Broekens, J., Heerink, M. and Rosendal, H. 2009. Assistive social robots in elderly care: a review. *Gerontechnology*, 8(2), 94–103.

Brooks, R. A. 2002. *Robot: The Future of Flesh and Machines*. London: Allen Lane.

Buckingham, R. and Graham, A. 2005. Snaking around in a nuclear jungle. *Industrial Robot*, 32(2), 120–127.

Dix, A., Finlay, J., Abowd, G. and Beale, R. 2004. *Human-Computer Interaction*. 3rd Edition. Boston, MA: Pearson Education.

Dodge, M. and Kitchin, R. 2009. Software, objects, and home space. *Environment and Planning A*, 41, 1344–1365.

Dourish, P. 2004. *Where the Action Is: The Foundations of Embodied Interaction*. London: MIT Press.

Dreyfus, H. 2000. Telepistemology: Descartes' Last Stand, in *The Robot in the Garden: Telerobotics and Telepistemology in the Age of the Internet*, edited by K. Goldberg. London: MIT Press, 48–63.

Fildes, J. 2006. Snake-arm robots slither forward. *BBC News*. [Online]. Available at: http://news.bbc.co.uk/1/hi/technology/5324708.stm [accessed 1 July 2008).

Fox, T. 2009. It's All Relative: UC San Diego's Einstein Robot Has 'Emotional Intelligence'. *UC San Diego News*. [Online 13 February]. Available at: http://ucsdnews.ucsd.edu/newsrel/science/02–09EinsteinRobot.asp [accessed: 19 January 2010].

Freud, S. 1985. The Uncanny, in *Penguin Freud Library Volume 14: Art and Literature*, edited by J. Strachey. London: Penguin, 339–376.

Gates, B. 2007. A Robot in Every Home. *Scientific American*, 296(1), 58–65.

Greenberg, J. 2008. The Healthcare Revolution. *JapanInc.com*. [Online 3 July]. Available at: www.japaninc.com/ [accessed: 15 August 2009].

Gross, D. 2009. Why Japan Isn't Rising', *Newsweek*. [Online 27 July]. Available at: http://www.newsweek.com/id/207063 [accessed 15 August 2009].

Guizzo, E. 2008. 6.5 Million Robots Now Inhabit the Earth. *IEEE Spectrum*. [online]. Available at http://spectrum.ieee.org/blog/robotics/ [accessed 15 August 2009].

Hagist, C. and Kotlikoff, L. J. 2006. Health Care Spending: What the Future Will Look Like. *NCPA Policy Report*, 286. [Online June]. Available at: www.ncpa.org/pub/st/st286 [accessed 15 August 2009].

Heim, M. 1993. *The Metaphysics of Virtual Reality*. Oxford: Oxford University Press.

Henig, R. M. 2007. The Real Transformers, *New York Times Magazine*, 29 July, 29–55.

Holz, T. R., Dragone, M., O'Hare, G. M. P. 2009. Where Robots and Virtual Agents Meet: A Survey of Social Interaction Research across Milgram's Reality-Virtuality Continuum. *International Journal of Social Robotics*, 1(1), 83–93.

Inievski, K. 2007. *Wireless technologies: circuits, systems, and devices*. Boca Raton, FL: CRC Press.

iRobot 2009. Our History. [Online]. Available at: http://www.irobot.com/sp.cfm?pageid=203 [accessed: 15 August 2009)

Jentsch, E. 1995. On the Psychology of the Uncanny. *Angelaki*, 2(1), 7–16.

Kidd, C., Taggart, W. and Turkle, S. 2006. *A sociable robot to encourage social interaction among the elderly*. Paper to the IEEE International Conference on Robotics and Automation (ICRA), Orlando, USA, May 2006.

Kiesler, S. and Hinds, P. 2004. Introduction: Special Issue on Human-Robot Interaction, *Human-Computer Interaction*, 19(1), 1–8.

Lee. K. 2004. Presence, Explicated. *Communication Theory*, 14 (1), 27–50.

MacDorman, K.F. and Ishiguro, H. 2006. The uncanny advantage of using androids in cognitive and social science research. *Interaction Studies*, 7(3), 297–337.

Melanson, D. 2009. Human-carrying robot bear gets cuteness upgrade. *Engadget. com* [Online 27 August] Available at: www.engadget.com [accessed 27 August 2009].

Mori, M. 1970. The Uncanny Valley. *Energy*, 7(4), 33–35 [Online]. Available at: androidscience.com [accessed 15 August 2009]

Myers, B. A. 1998. A Brief History of Human Computer Interaction Technology. *ACM Interactions*, 5(2), 44–54.

Palmer, J. 2008. Smart future for swarming robots. *BBC News*. [Online]. Available at: http://news.bbc.co.uk/1/hi/technology/7549059.stm [accessed 1 July 2008].

Paterson, M. 2006. Feel the Presence: The Technologies of Touch. *Environment and Planning D: Society and Space*, 24(5), 691–708.

Paterson, M. 2007. *The Senses of Touch: Haptics, Affects and Technologies.* Oxford: Berg.

Pineau J., Montemerlo M., Pollack M., Roy N. and Thrun S. 2003. Towards robotic assistants in nursing homes: Challenges and results. *Robotics and Autonomous Systems*, 42, 271–281.

Rutherford, M. 2008. Snake-arm robot works in tight quarters. *CNet.com* [online] Available at: http://news.cnet.com/8301–13639_3–10019958–42.html [accessed 15 August 2009].

Sheridan, T. B. 1992. Musings on telepresence and virtual presence. *Presence: Teleoperators and Virtual Environments*, 1, 120–126.

Shneiderman, B. 1987. *Designing The User Interface: Strategies For Effective Human-Computer Interaction.* Reading, MA: Addison-Wesley.

Steuer, J. 1992. Defining virtual reality: dimensions determining telepresence. *Journal of Communication*, 42(4), 73– 93.

Thrift, N. 2003. Closer to the machine? Intelligent environments, new forms of possession and the rise of the supertoy. *Cultural Geographies*, 10, 389–407.

Thrift, N. 2008. *Knowing Capitalism.* London: Sage

UN and IFFR 2002. *United Nations and the International Federation for Robotics: World Robotics 2002.* New York: United Nations.

Wada, K. and Shibata, T. 2008. Social and physiological influences of living with seal robots in an elderly care house for two months. *Gerontechnology*, 7(2), 235.

Weschler, L. 2006. Why Is This Man Smiling? *Wired Magazine*, 10.06, 16–17.

Wood, G. 2003. *Edison's Eve: A Magical History of the Quest for Mechanical Life.* New York: Anchor Books.

Woods, S. 2006. Exploring the design space of robots: Children's perspectives. *Interacting with Computers*, 18, 1390–1418.

Chapter 7

Technology and Good Dementia Care: An Argument for an Ethics-in-Practice Approach

Hilde Thygesen and Ingunn Moser

Introduction

Mr. Edvards broke his hip two years ago, and since then he has been very unsteady on his feet. In order to avoid falling he is dependent upon assistance when walking, either using his walking frame or needing help from one of the carers. At night time he usually gets up in order to go to the toilet, and although the night carer makes sure that his walking frame is positioned right next to his bed, he tends to forget to use it. This means that he is very unsafe and the carers are concerned that he might fall on his way to the toilet.

For some time now Mrs. Knutsen has been up at night a lot, rummaging around her room. Moreover, during the past few weeks, the carers have found urine on the floor on a regular basis. This is considered a problem as Mrs. Knutsen has slipped twice on the wet floor and fallen. The carers are worried that she may fall again and injure herself.

The above stories serve to situate this chapter in the everyday challenges and practices of dementia care. The topic of this chapter is thus dementia care, the ubiquitous ethical dilemmas in these care practices, and the role of technologies and other material arrangements therein. The question we explore is what constitutes good dementia care: what values and objectives that characterize good dementia care, but also how these are achieved and handled in practice, and realized in and through various material arrangements of care practice. In order to address issues of good care we build on a tradition of empirical ethics as developed by Jeannette Pols (2007) and Annemarie Mol (2008). This tradition approaches ethics from *within* care practices. Whether care is good is not judged from the outside, based on abstract ideas of what good care should be, as in traditional (bio)ethics, but engages instead the situated ideals, limits and reflections in care practice. Further, as much of the public debate on dementia care in Norway (and elsewhere) has been focused on bad care and care failures (Thygesen 2009), we have here chosen to focus on good care and highlight the ethics-in-practice that underlies this. For this purpose we work through empirical data from ethnographic observation of dementia care practices, and mobilize analytical tools and theoretical resources from a body of work on health, medicine and care in Science and Technology Studies (STS).

It is thus a concern underlying this chapter to acknowledge and appreciate the significance of the seemingly trivial and often mundane work of everyday dementia care.[1] This work is often unrecognized and ignored by bioethical and policy discourse and consequently experienced by carers as unrelated to ethics. Our argument is that good dementia care may be understood as a creative, ongoing process of trying out and assessing different care arrangements. This process is ethical in the sense that it involves a number of different values that the carers strive to realize and balance in practice. The values that are involved and how they are weighted is not given. Good dementia care is about what we will call 'sustaining the person'. The content and meaning of this notion will be developed throughout this chapter.

In addressing issues of personhood we choose to treat 'the person' as emerging and manifesting her- or himself through positions in practices and interactions with others. This stands in opposition to an understanding of personhood as some innate capacity or essence expressing itself. Put differently, we could say that we are not *born* persons, we *become* persons through the subject positions we are offered, take up, and become inserted into, in engagement with others – but also through positions we are denied, do not take up, and never enter (Moser 2003: 31). In this, we follow a semiotic and post-structuralist tradition which speaks of subjects and subject positions, and understands 'the subject' in the singular as composed of and drawing together a more complex set of different positions in which subjectifying capacities are manifested.

The subject in this usage is a position in a discourse, a position in which one expresses a capacity to speak, and recognizes oneself and is recognized by others as an 'I' that speaks, thinks, reasons, knows, feels and/or is conscious. Following a materially sensitive semiotic tradition in Science and Technology Studies, this treats subjectivity as relational, but it also extends the relations one traces subjectivity in from discursive to material in the widest sense. Material environments, including technologies, also condition what kinds of engagements and capacities that can be demonstrated (Moser and Law 1999, 2003). Subject positions arise in practices, and are enabled, embodied, and shaped in particular ways in local settings. They rely on arrangements, they have to be arranged for, and they take work and effort.

A subject position is thus not a position one has or is structured into once and for all, but rather a set of differently structured positions one moves between and is moved through, in practices. Some of these positions, and the subjective capacities

---

1   Jeannette Pols uses the term 'care as usual' and 'necessary care' to denote everyday care tasks and routines (2004, 100).

and identities they attribute to us, we may desire, willingly embrace and include in our image of ourselves. Others, we refuse and deny.

Accordingly, when we speak about care as aiming to sustain the person, this implies a relational understanding of personhood or subjectivity, as well as the care relation. The person is sustained through care practices that enable her or him to achieve, develop and/or maintain a sense of self, of an 'I', understood as a position that is also attributed subjectivity or personhood by others. As these capacities and characteristics that make up the subject or person are enacted and re-enacted in relations with other actors and entities, it follows that these capacities and characteristics are not absolute, but enabled to a greater or lesser degree in the concrete settings and relations in which they are expressed. This also applies to capacities and ideals such as autonomy, which in this approach are better understood as potentials and ideals which have to be realized relationally.[2]

By employing the notion of 'sustaining the person' we thus move away from an understanding of care as implying a notion of a person as autonomous, rational and independent, to allow for multiple and contrasting values to co-exist in the enactment of what a person is and what makes a person. Further, the notion of 'sustaining persons' implies a collective rather than an individualized focus. Care is not just about meeting individual needs, but involves a large network of humans and technologies involved in the shared daily life and in the caring process.[3]

The notion of 'arrangements' is also central in this chapter. Our use of the notion is closely tied up with the term 'conditions of possibility', which has been deployed by Michel Foucault to trace empirically how discourses structure what it becomes possible to know, and how it can be known within certain historical periods. One of Foucault's aims was to show how this was done in local, situated practices and settings, and what actors, techniques and procedures were involved in this event in which knowing took place and realities emerged. These sets of material relations and interactions were called 'dispositifs', and denote an ordered arrangement through which facts, entities and realities emerge (1979, 1981. See also Moser 2003 and Moreira 2004). The argument then is that specific ordered arrangements set the conditions for practices and for what kind of realities that are made possible. Similarly, specific ordered arrangements of procedures, legal

---

2 The understanding of autonomy in relational terms is an issue that is explored in different bodies of literature. One such body of literature is 'ethics of care'. See for example Verkerk 1999 or Tronto 1993. Science Studies represents another body of literature in this field. See for example Winance 2006; Stollmeijer 2000 and Pols 2004, 2007.

3 Moser (2010) has shown that the framing of the person with dementia in terms of the care-collective has important implications for how the dementia illness is localized. Within the context of everyday care, the dementia illness is not primarily understood as something that is located in an individual brain, but in interactions and daily life. The dementia illness is hence understood as something that affects how people live together. In this way the person with dementia is also understood as 'collective'. See also Schillmeier 2009; Mol, Moser and Pols 2010; Kaufman 2006 and Taylor 2008.

regulations, technologies and humans set the conditions for care practices; for what kind of care that is made possible, as well as what kind of positions that are open to those who receive care and those who give care.

The concept of 'dispositif', especially in its translation as 'apparatus', has however been criticized because it appears too static and structuralist (Barry 2001). Within STS, the notion of 'actor-network' was partly introduced to avoid this problem. This concept was coined to make it possible to trace the dynamic networking activities that link together places, facts, artifacts and realities, and so may be said to structure them. There is however also other metaphors for that which 'makes do', intended to replace the static, apparatus-like 'dispositif'. One of them is John Law's use of the notion of 'assemblages'. The term 'assemblage' is about how things hold together without an external framework in what Law describes as a 'tentative and hesitant unfolding' (2005).[4] In this chapter, we follow Law in his emphasis on arrangements as active and evolving practices or processes that are re-worked, rather than as a static structure (ibid: 41). Further, we understand arrangements as heterogeneous, in the sense that they are both discursive and material.

The empirical basis of the chapter is data drawn from fieldwork in four different care homes for people with dementia in the Norwegian context.[5] These care homes were organized as group homes, with common areas as well as the resident's individual rooms. Each of the care homes also had permanent staff. The size of the care homes varied from five to twenty residents.

Fall prevention was an important and non-contested goal of care in each of the care homes. Fall prevention is also a problem in elderly care in general.[6] By connecting the empirical analysis to one particular set of situations, our objective is to bring out the details of how care works in practice. These details are of central importance, as they show that even minor changes or adjustments in care arrangements can have great implications for whether the person is sustained. In addition, a focus on details gives prominence to the efforts and ongoing work that is involved in trying out and maintaining different arrangements.

We proceed as follows: First we demonstrate how dementia care is assembled as a collective process of trying out different care arrangements. The objective is to highlight the creative elements of this process as well as the work involved.

---

4   Other authors who have addressed the notion of 'assemblages' are Verran 2001; Mol and Law 2004; and Moreira 2004.

5   The data were collected as a part of an extensive study of the use of smart home technology in dementia care in Norway (Thygesen 2009). The fieldwork included ethnographic observation, formal and informal interviews, as well as document analysis, and took place over a period of approximately two years from late 2003 until mid 2005. In addition a total of thirty-six interviews of key actors were conducted within the same time-span.

6   Falls among the elderly is a targeted research-area. According to statistics presented by the US Centre for Disease Control and Prevention does one in three adults above the age of 65 year fall each year. Of those who fall, 20–30 per cent suffer moderate to severe injuries that increase their chances of early death. In 2000 the costs of falls among the elderly in the US was estimated to more than 19 billion USD (See www.cdc.gov/ncipc/factsheets/fallscost.html).

Next, we focus on dementia care as an ethical practice, and show how the different care arrangements enact different values, which have to be weighed and balanced against each other, and so how the process of trying out different care arrangements is about ethics-in-practice (Mol 2002, 2008, and Moser 2003).[7] Finally, we develop the argument that ethics-in-practice is about sustaining the person, and briefly address the importance of context in finding an arrangement that sustains the person.

Dementia Care as a Creative Process

Our first aim is to explore how good care is constituted. We will examine this issue through an elaboration of the stories of Mr. Edvards and Mrs. Knutsen, as presented at the beginning of this chapter. The stories of Mr. Edvards and Mrs. Knutsen have been chosen because they give a broad and comprehensive picture of how fall prevention is addressed in dementia care in Norway. The fact that these cases represent two different material environments is important as it highlights how dementia care is different in different contexts.

Fall Prevention as Arrangements

Firstly, in going back to the stories of Mr. Edvards and Mrs. Knutsen it is clear that they describe very common situations in dementia care, where falls may have potentially devastating consequences. As said, preventing falls from happening is an important aim of care.[8] The question here is: how are falls prevented? The story of Mr. Edvards shows that this involves specific care arrangements, such as the positioning of the walking frame next to his bed. Every night, as part of the evening routine, a carer makes sure the walking frame is in its place. Further, these arrangements are materially heterogeneous as they involve both humans and technologies (Latour 1987). The issue we will explore in the following is what the other characteristics of these arrangements are, and how they are determined.

Arrangements are Temporary

In order to do so we will continue the story of Mr. Edvards:

> First Mr. Edvards' medication was changed, as the sleeping tablet he had been
> taking for some years was thought to make him dizzy and more likely to fall.

---

7   This approach to ethics-in-practice builds on a French tradition drawing on the seminar-work of Canguilhem (1943) and is more recently developed by Boltanski and Thevenot (1991).

8   According to Tinetti et al. 1994 and Kannus et al. 1999 are a great majority of fatal head injuries in elderly people related to falls.

Instead, he was given something to calm him down earlier in the evening. However, this did not seem to have any effect at all, the carers concluded after an observation period of a few weeks. He was still as unsteady as before. Moreover, this observation period involved the night carer going frequently to his room at night, which clearly disturbed Mr. Edvards. Being a light sleeper, he woke up practically every time the night carer entered the room. So an arrangement was needed that did not involve a carer going into his room to check on him on a regular basis, the carers agreed. Other possible arrangements, such as putting up the bed rail were discussed. However, the carers decided against this, as Mr. Edvards was considered to be likely to climb over it, putting him at an even greater risk of falling.

To start with, this narrative shows that there are several sets of care-arrangements involved in preventing Mr. Edvards from falling at night. In addition to the positioning of the walking frame next to his bed, his medication is changed. Then an arrangement of the night carer going frequently into his room is tried out. But there is even more going on. The carers continually monitor Mr. Edvards and discuss different options, such as putting up the rail on his bed. The aim of trying out these different arrangements is to prevent falls without disturbing Mr. Edvards. When changing the medication fails to improve Mr. Edvards' unsteady gait, other arrangements are tried out.

Our data shows that arrangements in dementia care are seldom permanent or fixed, but temporary. There are constant changes or shifts of different arrangements being tested out or rejected, such as the carer going into Mr. Edvards' room at night. Arrangements are, then, ongoing processes that require continuous effort.

The constant work involved in care is beautifully described in Annemarie Mol and John Law's article on diabetes care, 'Embodied Action, Enacted Bodies: The Example of Hypoglycaemia' (2004). Here Mol and Law show how maintaining a good blood sugar balance is a continuous care process that involves practices such as measuring, eating, observation and injecting (insulin). The point is that maintaining a stable blood sugar level or a steady gait is continuous work; it is an ongoing and active achievement.[9] When eating (or not eating) fails to improve the blood sugar balance, other arrangements, such as injecting insulin, may be tried. In our case, changing the medication did not help, and then one tried checking in on Mr. Edvards.

Arrangements Involve Trial and Error

But, what kind of work is involved in finding good arrangements for preventing falls at night? We will examine this question through the continuation of Mrs. Knutsen's story:

---

9   This is also the argument of Schillmeier and Heinlein 2009.

At first, when the carers observed urine on the floor they assumed that Mrs. Knutsen was having problems finding her way to the toilet. As verbal communication was very limited, the carers looked for non-verbal cues. The urine on the floor on a frequent basis was one such cue. And as Mrs. Knutsen had slipped and fallen on the wet floor, the carers agreed that something had to be done in order to prevent her from urinating on the floor. First a toilet chair was tried out. It was positioned right next to Mrs. Knutsen's bed, which was expected to provide her with the immediate visual cue of a toilet. Would it solve the problem of her finding the toilet at night? It didn't. The next few weeks showed that the toilet chair was hardly ever used, and the behaviour of urinating on the floor continued.

Going inside Mrs. Knutsen's room to see to her at night was not an option, as she is an extremely light sleeper. Usually, just opening the door to her room will wake her up, and once she is awake she is not likely to fall asleep again, the carers told me. So instead of going into her room to see if she was in bed, the night carer often stood at the door listening for any sounds, before deciding whether to enter. However, this arrangement was unsatisfactory, as it did not give very reliable information about what was going on inside Mrs. Knutsen's room. So the search for a good arrangement continued.

The story of Mrs. Knutsen demonstrates that the process of finding a good solution involves testing out different constellations of arrangements. Arrangements are thus about trials and errors, in the sense that the action keeps moving backwards and forwards for as long as necessary. So as the toilet chair is tried out, and the carers keep finding urine on the floor, other arrangements are tried out, such as listening at Mrs. Knutsen's door at night. The carers continually monitor and evaluate the different arrangements, making necessary changes or adjustments. The trial and error process is unpredictable in the sense that it is never given what arrangement(s) will prove to be good, or for how long. The carers hope the arrangement with the toilet chair will solve the pattern of urinating on the floor and hence prevent Mrs. Knutsen from falling at night. However, as Mrs. Knutsen does not use the toilet chair, this arrangement fails. Failure is inevitably an integral element of arrangements as trial and error, of the continuous and unpredictable process of care. This is also an important argument of Mol and Law (2004: 57) and Mol (2008).

It is, however, important to note that the search for a good arrangement does not start from scratch. As the carers have extensive experience in preventing falls at night, they have a repertoire of arrangements to draw from. And as the observations and experiences related to Mrs. Knutsen are discussed among the carers, different possible solutions are contemplated. In the case of Mrs. Knutsen, the carers start by mobilizing an arrangement that is already at hand; the toilet chair, and by changing their routines of going into Mrs. Knutsen's room at night.

An important aspect of the trial and error process of trying out different arrangements for Mrs. Knutsen is the limited verbal communication. This makes the work of finding a good arrangement particularly demanding and time-consuming. Autonomy as informed consent becomes irrelevant in this context, since it presupposes a capacity for rational decision-making and of being able to verbally communicate one's preferences. And as Mrs. Knutsen doesn't have this capacity, the carers have to rely on careful observations of non-verbal reactions and visual cues to find out whether an arrangement is working.

Arrangements as an Ongoing Accomplishment

This means that there are no ready-made solutions, and there is a lot of effort involved. But the question remains: *what does that mean – in practice?* The story of Mr. Edvards continues:

> The carers decided to try out an alarm cord solution. The arrangement that was chosen for Mr. Edvards was a slight modification of an already existing solution. There was an alarm cord plugged into the wall beside his bed. When the alarm cord is pulled, an alarm is transmitted to the night carer's telephone. However, Mr. Edvards had never used this alarm cord, and the carers considered it to be unlikely that he would remember to pull it before getting out of bed at night. So, in order to ensure that the cord was pulled when Mr. Edvards was leaving his bed, the alarm cord was attached to his pyjamas top with a safety pin. In principle this meant that every time he moved out of bed, the alarm cord would be pulled, alerting the night carer, who would come and assist him to the toilet.

This arrangement did work well for some weeks. Every time he got out of bed, the cord was pulled and an alarm was activated, making it possible for the night carer to come to his room and assist him to the toilet. However, during the following weeks it became obvious to the carers that this cord being stuck to his pyjamas seemed to annoy Mr. Edvards. The carers discovered that he had ripped several of his pyjamas tops in order to get loose of the alarm cord. And on at least one occasion he had even undressed to avoid the alarm cord altogether. Another problem with this arrangement had to do with the plug that fastens the alarm cord to the wall. If the plug loosens, the alarm will not work. And on a couple of occasions the carers found that this was the case: the plug had not been in place.

And then, one early morning, Mr. Edvards slipped and fell on his way to the toilet. No alarm was transmitted to the night carer's telephone, so he was left lying on the floor until the day shift carer came to see to him. And as Mr. Edvards was sent to hospital with a broken hip, the carers tried to get an overview of what happened. Two weeks later, Mr. Edvards was returned to the care flat, frail and bed-ridden. The carers continued to discuss how to prevent him from falling when and if he recovered. However, Mr. Edvards died a few weeks later, still bed-ridden.

This narrative shows that the work involved in finding a good arrangement does not stop until Mr. Edvards dies. Arrangements are, then, *an ongoing accomplishment*. This understanding clashes with the well-defined structures of implementation as described in official guidelines on the assessment and use of technical aids in care, which portray implementation as involving certain pre-set stages, with a movement in a given direction.[10] First a problem is identified, and then an assessment follows. Further, a solution is found and the problem is solved. After a period of time the solution is evaluated and the implementation is complete.[11] However, organizing an arrangement in practice is not a task with a clear beginning and a well-defined end. Instead it is a continuous process of changes, adjustments, fine-tuning and monitoring. It is a process that Annemarie Mol (2008) has named 'doctoring', which suggests that the activity involves the attuning of a number of complex configurations (or variables) to each other in an ongoing process. The point is that although everything may fit and seem to work well for a time, it may well slip again. Care is not about implementation, Mol points out, but about being attentive, inventive and persistent.

This means that initially, although the safety pin solution seemed to work, the carers' work of adjusting and monitoring the arrangement continued: Does Mr. Edvards appear to find the arrangement acceptable? Is the alarm activated every time he gets out of bed? And is the night carer able to get to his room in time to assist him to the toilet? These are all central issues and questions that the carers pay attention to on a daily basis. In other words, the focus is on *doing*: on what has to be done. Good care is about figuring out how the various actors concerned should collaborate to best improve the person's situation (Mol 2008: 56).

Arrangements: Failures, Tensions and Conflicts

Thus far, we have presented the work of trying out different care arrangements as if this were a process without tensions and conflicts. However, this picture needs to be nuanced, for tensions and conflicts are a prominent part of the care-process. At the care homes where these stories are taken from, for example, there were constant tensions and conflicts between individual and collective needs. In the case of Mr. Edvards, when routines were changed and the night carer went to his room frequently to see if he was awake and getting out of bed, a consequence was that less time was available for attending to the other residents' individual needs. Finding a good arrangement, then, is about juggling and attending to individual *and* collective needs. And these needs are often in tension.

Another form of tension was the conflict between the carers. Although the carers discussed the different care arrangements on a daily basis and usually agreed on a course of action, there were also regular disagreements about what

10   See for example 'Hjelpemiddelformidling – en del av et større system' [The allocation of technical aids. A part of a large-scale system] Rikstrygdeverket 2000.

11   These are the standards elements involved in the process of providing technical aids to the disabled and the elderly, a procedure that is standardised by the National Insurance Agency (Rikstrygdeverket). See for example 'Hjelpemidler og etikk' [Ethics and technical aids] 2001.

care arrangement to try out next, or whether a care arrangement was considered to be working successfully. This means that the work of trying out different arrangements is about finding workable compromises.

Other forms of tensions and conflicts were also part of the care process at the care homes, such as conflicts between the different needs of the person. Mr. Edvards' need for a good night's sleep, for example, was clearly in conflict with the arrangement of going into his room frequently at night in order to prevent him from falling.

Mol and Law (2004) maintain that dealing with these tensions is not about making rational choices between different clearly defined options. The ideal of perfect balance (such as between individual and collective needs) or harmony (in professional judgement or in meeting different individual needs) is not sustainable. The argument is that things don't add up. There are always some variables missing – they may behave unpredictably or are not known, such as the alarm plug falling out of its socket, which caused the collapse of the safety pin solution for Mr. Edvards. No matter how hard you try, unexpected failure may be the result.

Summing up: Arrangements as Creative Processes

In summing up this section, we argue that the examples of fall prevention demonstrate that good dementia care can be understood as a creative and collective process of trying out different care arrangements. The work of organizing the different arrangements is creative, in the sense that it is about attuning the many variables to each other in an ongoing process. It is about attending to what is needed, both at an individual level and collectively. This is not without difficulties, since there are always frictions and tensions. Even so, as Annemarie Mol points out, the logic of care requires us to keep on specifying: 'try, be attentive to what happens, adapt this, that or the other, and try again' (2008: 53). The success of this work is first and foremost dependent on the carers' ability to experiment with what can be done and to respond to the ever emerging frictions.

Good care is hence not a rationalist endeavour in the sense that is not merely or primarily a cognitive task that involves priorities, means and procedure rationality. Instead, it is about a different kind of rationality based on trial and error. It is about the endless work of monitoring, observation and fine-tuning. It is a creative and unpredictable process that involves a number of compromises.

Dementia Care as an Ethical Practice

What we have argued so far, is that good dementia care is about the creative work – or processes – of organizing different care arrangements. But there is more to it. We will now take the analysis further by addressing the issue of values in dementia care. What makes a good (or less good) care arrangement? Or, in other words: how to think about dementia care as an ethical endeavour? It is important to note that the notion of 'ethics' is here used in a broad sense; as about 'trying to do good'.

Each Arrangement Involves the (Creative) Balancing of Goods and Bads

The above stories of fall prevention show that while some arrangements are seen to work, others do not.

In the case of Mr. Edvards, the safety pin arrangement worked initially, as the night carer was alerted every time he got out of bed, and as Mr. Edvards does not seem to object to the limitations imposed on him by having the alarm cord fastened to his pyjamas top. However, Mr. Edvards increasingly found the alarm cord being pinned to his pyjamas top annoying and limiting, and the good safety pin solution became a bad one. Another important variable was the alarm plug that kept falling out of its socket, making the arrangement unstable and unreliable. What is a good (or bad) arrangement is hence not an absolute or given category, but a matter of care creatively attuning to the many variables that makes up the arrangement, and that may change over time. The conclusion to be drawn is that arrangements have both limiting and enabling aspects, which means that the process of trying out different care arrangements is also a way of creatively weighing and balancing goods and bads.

The trope 'weighing and balancing' is important, as it indicates that judgements are being made. However, it is important to note that these judgements are not definitive in form. This is not an academic discipline of reflecting systematically on problems of good and bad. Instead, weighing and balancing is used here to denote the active and creative process of attuning, of ethics-in-practice.

Bads Are a Necessary Part of All Arrangements

How is this weighing and balancing of goods and bads done in practice? Here is a continuation of the narrative about Mrs. Knutsen.

> The carers discussed different options, including the use of a mattress alarm. It was no coincidence that a mattress alarm was considered, as this technical aid was already in stock, having been purchased by the municipality when the care flat was first opened nearly three years ago. The mattress alarms had not been used, as the local dementia team considered them to be surveillance technology, since the alarm would detect any movements in and out of bed.[12]

---

12   The local dementia team is a multi disciplinary team appointed by the municipality to give professional advice on matters concerning dementia care.

A mattress alarm is a thin mattress like sensor that is placed in the bed. The sensor detects any movements and transmits an alarm through the smart home system to the night carer's telephone. The use of the mattress alarms was thereby restricted and subject to formal approval by the local dementia team, following the submission of a written application. These procedures were based on official guidelines.[13] So the carers knew that the mattress alarm would imply an infringement on Mrs. Knutsen's privacy, since the night carer would know whether she was in bed at all times during the night. Yet they also knew that this alarm might increase the likelihood that the night carer would be able to assist Mrs. Knutsen to the toilet, reducing the risk of her urinating on the floor. So, it was worth trying, they decided.

But first the necessary paperwork had to be done. In the written application the carers had to describe in detail Mrs. Knutsen's problems and the issues that the mattress alarm was expected to solve. The form was signed by the nurse in charge, as well as by Mrs. Knutsen's son, as her next-of-kin. The application was not signed by Mrs. Knutsen as she was considered by the nurse in charge not to be competent of giving her informed consent. So her son was asked to sign on her behalf. This was done despite the fact that Mrs. Knutsen was not formally declared incompetent. Very few people with dementia are formally declared as incompetent in Norway, and it is a common procedure to get the next-of-kin to give their informed consent in these kinds of matters. This practice is also legally approved.

The application was approved by the dementia team within a couple of days. Since the alarm was already in stock, all that was needed was for the carers to call the caretaker, and ask him to come and install it.

The carers were uncertain. Would the alarm prove to be a good solution? The alarm was tried out. And yes, it did seem to work. Now the night carer was alerted every time Mrs. Knutsen was in the process of moving out of her bed. Usually she was still sitting on the side of her bed when the night carer arrived to assist her to the toilet and back. The pattern of urinating on the floor stopped. And the night carer no longer had to open Mrs. Knutsen's door at night to know whether she was in bed.

This is a story of a good arrangement – of an arrangement that worked to provide good care for Mrs. Knutsen, at least for the time being. The arrangement of the mattress alarm allowed information of Mrs. Knutsen's movements to be transmitted to the night carer's telephone. This in turn made it possible for her to know when Mrs. Knutsen was moving out of bed, and hence to assist her to the toilet.

---

13   'Hjelpemidler og etikk' [Technical aids and ethics], Rikstrygdeverket 2001 and 'Smart home technology. Planning and management in municipal services', Directorate for Health and Social affairs, The Delta Centre 2005.

Using a mattress alarm was considered to be better than waking her up. However, this arrangement also had its bads or limitations, as it allowed for the constant detection of Mrs. Knutsen's movements in bed. As these limitations are defined by ethical- and legal rules as an intrusion on her privacy, they are subject to certain procedures.[14] However, what we are proposing is that bads or limitations cannot be avoided; they are a necessary part of all arrangements. In the story of Mrs. Knutsen, the point is that it is *precisely* the detection of her movements in bed (the bads or limitations of the mattress alarm) that makes it possible for her to get assistance when she needs it without being disturbed in her sleep.

The resulting argument is that bads will have to be weighed in the total picture to find the best arrangement. In other words, the different arrangements must not be seen in isolation or out of context, but as a part of a whole. This approach conflicts with the understanding of certain technologies (and arrangements) as being bad in themselves and as separate from practices.[15]

Creative Ethics Involve Different Values

What we have shown is that arrangements are ethical in the sense that they involve the balancing (the striving for, realization and manifestation) of different values, including bads or limitations. But *what are these values?* Returning to the narratives of Mr. Edvards, we will highlight some of the values in the different arrangements that were involved in preventing him from falling at night.

In the walking frame arrangement, the night carer placed the walking frame next to Mr. Edvards' bed so that he would be able to use it when he got out of bed at night. However, the carers discovered that Mr. Edvards hardly ever used the walking frame despite it being placed right next to his bed. Although this arrangement preserved Mr. Edvards' autonomy, in the sense that he could decide for himself whether or not to use the walking frame, it was not a good arrangement. In this context, being autonomous was too limiting, as it did not secure the necessary support or assistance needed to prevent him from falling. In being made autonomous, the responsibility for preventing falls at night was made to rest upon Mr. Edvards. However, the fact that he did not use the walking frame is not necessary an expression of his own preferences and a conscious decision not to use the walking frame. It may equally have been a consequence

---

14 'Smart home technology. Planning and management in municipal services', Directorate of Health and Social Affairs, 2005: 26–29. 'Hjelpemidler og etikk' [Technical aids and ethics], Rikstrygdeverket 2001: 13 and 15.

15 Stortingsmelding [Report to the Storting] nr 28 (1999–2000): 62–63. This report specifically addresses the issue of technology and dementia care. Different types of technologies are classified according to how intrusive they are perceived to be. Five different categories of technologies are identified ranging from 'ordinary technologies and technical aids' to 'technical solutions that limits the personal freedom'. The point is that the technologies are considered as given outside the context of their use.

of his illness, as dementia involves the erosion of cognitive functions, including memory, orientation and the capacity for rational decision-making. Mr. Edvards may have forgotten to use the walking frame when he was getting out of bed at night. It is impossible to know. The point is that if the carers had assumed that not using the walking frame was an expression of Mr. Edvard's rational choice, and they had consequently failed to explore other alternatives, this would have been neglect, which is bad care.

The arrangement of the night carer going frequently into Mr. Edvards' room at night in order to see if he was getting out of bed, did make it possible for him to get the necessary assistance, i.e. the needed safety or protection from falling. However, this arrangement also had its limitations, as it disturbed his sleep. So the value of a good night's sleep was involved. Other values were also at stake, such as care, in the sense of being looked after and attended to, and privacy.

In the arrangement using the safety pin and the alarm cord, Mr. Edvards did initially get the necessary assistance. When he moved out of bed, the alarm cord was pulled and the night carer alerted. So also this arrangement involved the good of safety, or the necessary assistance and protection from falling. In addition it was not necessary for the night carer to come to his room frequently at night, which disturbed his sleep. So this arrangement also made the good of a good night's sleep possible. A major, but necessary limitation of this arrangement was, however, the physical attachment of the alarm cord to Mr. Edvards' pyjamas top, impinging on the value of freedom of movement.

An important point is that it is the *specific configurations* of the arrangements that enact the different goods and bads. The above analysis also shows that there are multiple goods in health-care. The values that are involved and how they are ranked are outcomes of the weighing and balancing of the specific arrangements. In the case of Mr. Edvards, undisturbed sleep, safety and freedom of movement were more important values than individual choice. This means that there is no simple or given hierarchy of values in dementia care.

The Aim of Ethics-in-Practice: 'Sustaining the Person'

Ethics-in-practice thus acknowledges that dementia care involves a number of different values and that their ranking is not given. Further, these values are manifested and realized through the creative process of trying out different care arrangements. A similar argument has been made by Pols (2004) and Mol (2008). In this section we will argue that this creative work of organizing arrangements by weighing and balancing different values is about sustaining the person.

The notion of 'sustaining the person' implies a relational understanding of care and the care relation, and is closely related to the understanding of 'relational autonomy' as associated with feminist bioethics and ethics-of-care (See for example MacKenzie and Stoljar 2000, Tronto 1993 and Verkerk 1999). Feminist bioethics and ethics-of-care share a concern with the contextual and relational. The argument is that the traditional bioethical ideal of autonomy as self-determination is forwarded at the expense of a recognition of the value of relations, of dependence and interconnection. Thus the shifting contexts in which the person is embedded is moved to the forefront of attention.

The notion of 'sustaining the person' as relational autonomy is based on an understanding of the self as relationally constituted and embedded. This means that individual autonomy is not perceived of as an innate capacity of humans in the humanist sense, but as *one* possible outcome of concrete relations. As noted, within a 'sustaining the person' approach, also the relations between humans and technology are emphasized: as these relations, equal to the relations between humans, set the conditions for what kind of capacities that are made possible and can be recognized. Autonomy is hence not a given capacity but rather an ideal and potential that needs to be realized, and also realized in relations. Whether these relations are good (or serve to sustain the person) is not given, but needs to be judged in the actual context of care.

However, in relating the notion of 'sustaining the person' to relational autonomy we are wary of placing too much emphasis on autonomy. So, although the notion of relational autonomy closely resembles that of sustaining the person, these two concepts are also distinct. The point we are making is that within a logic of sustaining the person, the care arrangements of Mr. Edvards and Mrs. Knutsen are not necessarily understood in terms of autonomy. Preventing Mr. Edvards and Mrs. Knutsen from falling at night involves a number of other important dimensions, such as a good night's sleep, freedom of movement, and also preservation of freedom of movement and integrity by preventing falls, fractures and injuries. And within an understanding of dementia care in terms of 'sustaining the person' these dimensions are given a central place.

An important element of the understanding of dementia care as 'sustaining the person' is its collective focus. Good care is here not just about taking care of Mr. Edvards and Mrs. Knutsen, but includes a much larger network of humans and technologies involved in the process of caring. Hence the aim of ethics-in-practice can equally be seen to be about sustaining the carer. Sustaining the person is about finding a way that makes it possible to get through the day in a way that is the least limiting for both carers and the cared for.

What is a Good Arrangement Has to be Decided in Context

We will briefly address the importance of context in finding an arrangement that sustains the person. In order to do so we will tell the story of Mrs. Solberg and her wooden bed side arrangement:

> As Mrs. Solberg kept falling out of bed at night, the carers mobilized a succession of different arrangements to prevent this. One such arrangement involved an ordinary bed rail. All of them failed.

Although Mrs. Solberg had no verbal language, it was clear that she was very anxious and distressed about the situation. So the carers discussed what to do. In the end they decided to ask the local caretaker if he could come up with a solution to the problem. And he did. A high bed guard was made of solid wood and attached to the bed. Finally, Mrs. Solberg was protected from falling out of bed. As a result, the carers agree, she has clearly become more content with the situation. She does not look as anxious as before, and she co-operates with the carers as they assist her with getting into bed.

The story shows that it is not given what arrangement(s) may work to sustain the person. This means that a wooden bed guard arrangement is not necessarily unethical in itself. On the contrary, the unconventional arrangement clearly makes Mrs. Solberg feel safe, as she is no longer afraid of falling out of bed. This solution does limit her movements, since she is unable to get out of bed on her own. But in this case, the limiting of her movements is precisely what makes it a good arrangement for Mrs. Solberg.

According to bioethical ideals, this kind of arrangement is considered unethical in itself, as it hinders Mrs. Solberg's freedom of movement and thereby her individual autonomy; she is not able to decide for herself when (or if) she wants to get out of bed. According to Norwegian legal rules these kinds of arrangements are illegal to use in most circumstances.[16] Other arrangements involving new technologies, such as the mattress alarm for Mrs. Knutsen, are seen to be problematic, as their use is considered as an intrusion on the person's individual autonomy. If arrangements that involve these types of technologies are contemplated, certain formal legal and ethical procedures are enforced to ensure that the legal rights (or individual autonomy) of the person is maintained. In practice this means that care arrangements involving these technologies are only to be used if a written informed consent is obtained, or in an emergency situation when patient autonomy is temporarily suspended.

A serious implication of following these principles is that a number of potentially good arrangements are ruled out on the basis of abstract/decontextualized principles alone. The consequence is that legal and ethical principles limit the ways in which the person may be sustained. However, our point is that there is another tradition which emphasizes the situatedness of care and its ethics. This is a tradition that has *practices* as a starting point. Our recommendation is that this understanding of ethics in terms of ethics-in-practice needs to be developed and strengthened.

Conclusion

In this chapter we have addressed issues of good dementia care. Through the mobilization of tools and resources from a body of literature on health, medicine

---

16   See for example Stortingsmelding (Report to the Storting) nr 28 (1999–2000): 63.

and care within the STS-field, as well as empirical data from studies of dementia care practices, we have shown that good dementia care can be understood as an ongoing and creative process of trying out different care arrangements. The process is creative in the sense that it is never given what arrangement(s) will turn out to be good or for how long. Thus care is about trial and error, and the neverending work of experimenting with what can be done, both at an individual level and collectively. As this process is filled with frictions and tensions, this work is difficult and demanding. This means that care is not a rationalist endeavour, in the sense that it is not simply a cognitive task that involves priorities, means and procedure rationality, which are stressed by bioethical ideals.

Further, we have argued that this process of trying out different care arrangements is an ethical practice, in the sense that it involves the weighing and balancing of different values. The values that are involved and how they are weighed is not given, but is an outcome of the specific configurations of the care arrangement(s). However, an important point is that all arrangements involve limitations, or bads. This means that the bads will have to be weighed in the total picture to find the best arrangement. In order to know whether or not an arrangement is good, it is necessary to see the arrangement in its context, as a part of a whole, and not in isolation. Through different stories of fall prevention, we have shown that good care is about sustaining the person. The arrangements that may work to sustain the person are not given, and have to be decided in context.

An important aim of this chapter has been to highlight and acknowledge the work and tremendous effort involved in providing good dementia care. The stories of Mr. Edvards and Mrs. Knutsen demonstrate the significance of everyday activities and routines, such as those involved in preventing falls at night. In this chapter we have shown that these practices are indeed important, as they are ethics-in-practice.

References

Barry, A. 2001. *Political Machines. Governing a Technological Society*. London and New York: The Athlone Press.

Boltanski, L. and Thevenot, L. 1991. *De la Justification: Les Economies de la Grandeur*. Paris: Gallimard.

Canguilhem, G. 1943. *Le Normal et le Pathologique*. Paris: PUF.

Foucault, M. 1979. *Discipline and Punish. The Birth of the Prison'*. Harmonsworth: Penguin.

Foucault, M. 1981. *The History of Sexuality. Volume 1: An Introduction*. Harmonsworth: Penguin.

Kannus, P., Parkkari, J., Koskinen, S. et al. 1999. Fall-induced Injuries and Deaths among older Adults'. *JAMA*, 281: 1895–1899.

Kaufman, S. 2006. Dementia-Near-Death and Life it-self. In *Thinking about Dementia. Culture, Loss and the Anthropology of Senility*, edited by A. Leibing and L. Cohen. New Brunswick: Rutgers University Press.

Latour, B. 1987. *Science in Action: How to Follow Scientists and Engineers through Society*. Milton Keynes: Open University Press.

Law, J. 2005. *After Method. Mess in Social Science Research*. London and New York: Routledge Taylor and Francis Group.

MacKenzie, C. and Stoljar, N. (eds). 2000. *Relational Autonomy. Feminist Perspectives on Autonomy, Agency and the Social Self*. New York and Oxford: Oxford University Press.

Mol, A. 2002. *The Body Multiple: Ontology in Medical Practice*. Durham, N. Ca., and London: Duke University Press.

Mol, A. 2008. *The Logic of Care. Health and the Problem of Patient Choice*. London and New York: Routledge Taylor and Francis Group,

Mol, A. and Law, J. 2004. Embodied Action, Enacted Bodies: The Example of Hypoglycaemia. *Body and Society*, 10 (2–3): 43–62.

Mol, A., Moser, I. and Pols, J. 2010. *Care in Practice: On Tinkering in Clinics, Homes and Farms*. Bielefeld: transcript Verlag.

Moreira, T. 2004. Self, Agency and the Surgical Collective: Detachment. *Sociology of health and Illness*, 26 (1): 32–49.

Moser, I. 2003. *Road Traffic Accidents: the Ordering of Subjects, Bodies and Disability*. Dr.art thesis, Faculty of Arts, University of Oslo, Norway.

Moser, I. 2010. Perhaps Tears Should not be Counted but Wiped away. On Quality and Improvement in Dementia Care. In *Care in practice: On Tinkering in Clinics, Homes and Farms*, edited by A. Mol, I. Moser and J. Pols. Bielefeld: Transcript Verlag.

Moser, I. and Law, J. 1999. Good Passages, Bad Passages. In *Actor Theory and After*. Edited by J. Law and J. Hassard. Oxford: Basil Blackwell, 196–219.

Moser, I. and Law, J. 2003. Making Voices: New Media Technologies, Disabilities and Articulation. In *Digital Media Revisited. Theoretical and Conceptual Innovation in Digital Domains*, edited by G. Liestöl, and T. Rasmussen. Cambridge: MIT-Press: 491–520.

Pols, J. 2004. *Good Care. Enacting a Complex Ideal in Long-term Psychiatry*. Ph-D Thesis, Trimbos-instituut, Utrecht, the Netherlands.

Pols, J. 2007. Which Empirical Research, Whose Ethics? Articulating Ideals in Long-term Mental Health Care. In *Empirical Ethics in Psychiatry*, edited by G. Widdershoven et al. Oxford: Oxford University Press.

Rikstrygdeverket (National Insurance Agency). 2002. *Hjelpemiddelformidling – en del av et større system* [The allocation of technical aids. A part of a large-scale system].

Rikstrygdeverket (National Insurance Agency). 2002. *Hjelpemidler og etikk* [Ethics and technical aids].

Schillmeier, M. 2009. Actor-networks of Dementia. In *Un/knowing Bodies*, edited by J. E. Latimer and M. Schillmeier. Blackwell Publishing Ltd/The Sociological Review, 139–158.

Schillmeier, M. and Heinlein, M. 2009. From House to Nursing Home and the (Un)Canniness of Being at Home. *Space and Culture*, 12(2): 218–231.

Sosial og helsedepartementet (The Ministry of Health and social affairs). 2000. *Stortingsmelding* [Report to the Storting] nr 28 (1999–2000) Innhald og kvalitet i omsorgstjenestene. Omsorg 2000.

Sosial og helsedirektoratet (The Directorate for health and social affairs). 2005. Smart home technology. Planning and management in municipal services', The Delta Centre.

Stollmeijer, A. 2000. *The Ethics of Care. Alzheimer's Disease and the Question of How to Give Good Care?* Paper at the 4S/EASST Conference: Worlds in Transition: Technoscience, Citizenship and Culture in the 21st Century, Vienna, September 27–30.

Taylor, J. 2008. On Recognition, Caring and Dementia. *Medical Anthropology Quarterly*, 22 (4): 313–335.

Thygesen, H. 2009. *Technology and Good Dementia Care. A Study of Technology and Ethics in Everyday Care Practice.* Ph-D thesis, centre for Technology, Innovation and Culture (TIK), Faculty of Social Sciences, University of Oslo.

Tinetti, M. E., Baker, D. I., McAvay et al. 1994. A Multifactorial Intervention to Reduce the Risk of Falling among the Elderly Living in the Community. *New Engl. J. Med*, 331: 821–827.

Tronto, J. C. 1993. *Moral Boundaries: A political Argument for an Ethic of Care.* New York: Routledge.

Verkerk, M. 1999. A Care Perspective on Coercion and Autonomy. *Bioethics*, 13, (3/4): 358–368.

Verran, H. 2001. *Science and an African Logic*. Chicago: University of Chicago Press.

Winance, M. 2006. Trying out the Wheelchair: the Mutual Shaping of People and Devices through Adjustment. *Science, Technology and Human Values*, 31: 52–72.

Chapter 8

Social Remembering as an Art of Living: Analysis of a 'Reminiscence Museum'

Elena Bendien, Steven D. Brown and Paula Reavey

Introduction

Karin is eighty-four years old. She lies still, her eyes open looking round the room she is in. She sees the table with places set for four people. A lace tablespread drapes across it, slightly awkwardly bunched at one end. She sees the sideboard radio. A tune from some fifty (or is it now sixty?) years ago drifts across the room. Cheap mass produced paintings adorn the walls. There are several pairs of socks drying on a rail around the fireplace. She struggles to sit up. Her mouth forms a word – 'home'.

But Karin is not at home. She is in the basement of a large elderly care home facility on the outskirts of Rotterdam in The Netherlands. Karin is lying on a hospital bed. She is in the final stages of advanced dementia and has not talked now for some time. In a few months she will pass away. Karin lives in the full time nursing care part of the home. Today she has been taken on a rare trip out of the ward to a new facility in the home, something known as a 'Reminiscence Museum' (*Herinneringsmuseum*). She has been wheeled on her bed into one of the rooms in the museum that has been arranged to resemble a typical Dutch living room from the 1930s/1940s. When she utters the word 'home', it is this peculiar artificial space to which she is referring.

The reminiscence museum consists of a series of specially constructed rooms built in the basement of the 'Akropolis' care home. Most of rooms have been carefully arranged to resemble a Dutch domestic space from the first half of the twentieth century. For instance, 'the nursery' is comprised of a bed, cabinets and a large collection of toys which are arranged to create the effect of a child's bedroom. All of the objects in the space can be picked up and inspected (visitors are in fact encouraged to do so by staff in the museum). The museum has amassed a huge range of artefacts, ranging from period furniture, decorations and clothing to kitchen equipment and toys. The items are for the most part donated or specially acquired by the curator, Inez van den Dobelsteen-Becker. Some rooms house large collections of key items (e.g. cameras) which are kept in glass cabinets. Others – such as the space known as 'the workshop' – are almost overwhelmed with objects which are almost heaped on top of one another. One room – 'the grocery shop' – stands out since it has been arranged to resemble a neighbourhood shop, filled with boxes of period grocery items, jars of sweets and a working cash register. This room was one of the earliest to be completed when work on the museum began in 2006.

In this chapter we will describe this museum as a very particular kind of care setting that redefines what is meant by 'care'. Indeed one of the most interesting features of the museum is that it is specifically intended to not be a care setting at all. Hans Becker, the director of care provider – Humanitas – who run the facility in which it is based, has stated that the purpose of the museum is simply to facilitate 'happiness'. It has nothing to do with care, at least in the formal medical sense of the term. Becker instead sees the museum as designed to support an 'art of living' which Humanitas aims to foster amongst its elderly clients.

If the reminiscence museum is not a formal care setting, then what exactly is it? We will use the term 'social technology', derived from recent work by Maarten Derksen and Signe Vikkelsø,[1] to characterise how the museum fits with the aims of Humanitas. We will argue that it is the joint construction of memories about past times that occurs through the interaction of visitors and staff which is critical to an understanding of how the museum functions. Although this interaction is thoroughly mediated by the museum space and the assembled objects, it is not reducible to them. In this sense the museum is a structured social space for the enactment of practices of remembering. And it is these practices which are in turn integral to how the 'art of living' might be accomplished.

Humanitas and the 'Art of Living' in Later Life

Humanitas is a care-provider organization based in the Netherlands who specialise in care for the elderly (they also provide some care for marginal persons – drug users, sex workers and the homeless – who they refer to as the 'young old').[2] The organisation has a nearly fifty year history of running care facilities principally in the Rotterdam area. It currently maintains over thirty care homes. A pivotal moment in the history of Humanitas was the appointment of Hans Becker as director in 1992. Becker initiated a change process which sought to entirely reshape how Humanitas went about delivering care. As Letiche (2008) describes it, this new vision of care was based on a simple premise:

> The director of Humanitas began the change process by posing the following question: Assume that you are in the last two or three years of your life. You are alone and often very lonely. What would give you enough pleasure to get you voluntarily out of bed in the morning? The answer did not come as a grand flash of revelation, but as lots of small ideas. These ideas included: an appointment

---

1   See a forthcoming special issue of journal *Theory and Psychology* edited by Derksen and Vikkelsø.

2   The best summaries in English of Humanitas' work are Becker (2003) and Letiche (2008). The description in this section draws heavily on the latter.

to set coffee for the neighbor, a primary school child who is coming for lunch, a plan to drink tea in the atrium. The answers all had to do with the social surroundings, in which one lives. It helps if there is lots of color and sunshine in the building and art and objets d'art are all about. A giant aviary or a 375,000 liter aquarium would help. (Letiche, 2008: 187)

Humanitas built a series of experimental spaces to realise this vision. These spaces are mixed (in fact so mixed that planning permission was initially difficult to obtain since the applications did not fit into existing local government categories). For example, the Akropolis building in which the reminiscence museum is based comprises two blocks of apartments facing towards one another across a huge central covered atrium. The atrium opens out onto a restaurant and bar complex, which provides high quality subsidised food and drink. This complex is augmented with a small supermarket, a hair-dressing salon and an internet café. Together these are known as the 'village square'. Visitors to the home pass by way of a 'petting zoo' which keeps a range of small animals. Once inside, visitors are confronted with a range of antique wheelchairs (all of which are usable) and space in which seemingly every inch is either brightly painted in vibrant colours or contains an artwork or decorative object (e.g. a huge sitting Buddha, masks, old musical instruments). The atrium is typically filled with activity as visitors and clients move around the various facilities. On occasion visitors may encounter less predictable events. We have observed indoor markets, the arrival of coachloads of visitors to the museum, and on one particularly memorable visit the construction of a life-size nativity scene attended by an entourage of specially dressed Sinterklass helpers. Sadly we missed the much remembered occasion on which the director celebrated his birthday by riding a camel through the building.

The aim of this space is to transform what could be a 'misery island' (Becker, 2003) into a place of continuous activity where elderly clients have continuous opportunities to focus their attention on things beyond their own immediate health. For example, wheelchairs are usually potent symbols of debilitation. At Akropolis they are first of all aesthetic objects, there as 'conversational pieces' and sources of ironic humour. Although there are the full range of healthcare facilities based on site at Akropolis (including a nursing home for clients entering into stages of severe debilitation), these are arranged on the periphery of the space rather than the centre. By integrating the homes into the surrounding area and by making them fully accessible to the local community as well as the families of resident clients, Humanitas aims to turn its care-spaces inside out. Visitors are drawn to Akropolis for many reasons other than obligation. The reminiscence museum is in this sense an extension of previous practice rather than an entirely new direction. It was initially conceived as another innovation which would involve residents in their own environment (the museum is maintained in part by clients) and would bring visitors to Akropolis. It has however over time become a significant attraction in its own right, with elderly and younger persons from across the Netherlands coming to Rotterdam specifically to visit.

The change programme at Humanitas has involved more than just alterations spatial design. Dramatic changes to the structure and definition of roles within the organization has been made. Primary care, for instance, is seen as something which the elderly can provide for one another. Professional intervention takes on an anticipatory function (Letiche 2008). Nurses and doctors are encouraged to listen to residents and anticipate emerging medical problems rather than focus on (expensive) emergency care. Dieticians are required not merely to meet dietary needs but to turn meals into pleasurable experiences. For example, there are regular sessions based in the museum where pre-war traditional Dutch dishes are served.

The reorganization of roles and responsibilities has been guided by what Becker (2003) refers to as four 'core values'. These comprise:

- *Being in control* – individual autonomy must be respected, even if it invites negative social judgements (e.g. being drunk at the bar). This is supported by 'hands-behind-the-back-care' where caregivers should resist the temptation to take over activities simply because they can accomplish them quicker.
- *Active participation* – clients are encouraged to retain control and involvement in their own daily care – e.g. cooking, cleaning, mending clothes – for as long as possible. The slogans 'use it or lose it' and 'too much care is as bad as too little care' are often invoked to remind caregivers to facilitate self-care as far as possible.
- *Extended family approach* – divisions between clients and staff are to be eroded. This can range from treating clients and their immediate family as 'experts' in their own care to providing clients with roles within the organisation itself (e.g. part-time curators in the museum)
- *Yes-Culture* – employed and voluntary staff at Humanitas are required to adopt a positive attitude to any request made at the home. These may mean entertaining non-standard or peculiar requests (e.g. to retain a number of pets).

Together these core values map out a version of care which differs markedly from a biomedical approach. In particular they emphasise the social relations between staff and clients as the primary site where care occurs. However unlike the classical model of care, where the client is dependent on the expertise and relative power of the care-provider, here efforts are made to invert that relationship. Care often happens precisely when the clinician refuses to automatically exercise expertise, instead encouraging and supporting the client in providing their own care and setting the agenda for what care might be.

Given this approach it is unsurprising that Becker (2003) often shies from the very word 'care' itself, preferring to use unspecified formulations such as 'making people happy' or 'let many flowers blossom'. The apparent vagueness of these terms belies the fact that the term 'care' is considered an object for continuous inspection at Humanitas, which has a standing research group of managers, staff and academics charged with investigating the meanings of care throughout the organisation (see Letiche 2008).[3] Becker's conception of an 'art of living' is a more formal drawing together of these many 'small ideas' which have been developed in response to the four core values, along with a range of larger ideas from the European Humanist tradition. Key amongst these is the notion of living as the process of giving meaning to existence, as an ongoing questioning of what it is to be living this particular life. This implies critical reflection in general and specifically, in the case of ageing, on the representations of age that structure the possibilities for growing old. Participation in social life, along with personal autonomy, are also central concepts – 'Both in the ethics of the art of living and in the Humanitas system, it is strongly emphasized that people, and therefore also older people, should as far as possible be the architects and designers of their own lives' (Becker 2003: 188).

The concept of an 'art of living' applied to later life clearly differs from the usual notion of 'cure' and 'care' as a guiding principles. But is also complicates the question of what precisely the purpose is of reminiscence museum. If its function is not that of providing care in the traditional sense, then what exactly does it do? Is it sufficient to describe and evaluate the museum using one of Becker's formulations – as a place where clients are 'made happy'? Or can it be seen as a support for an 'art of living'? What exactly might that then entail in relation to the museum?

Social Technology

We regard the reminiscence museum as a 'social technology'. The meaning of this peculiar term is worth spelling out a little. In his later work, Michel Foucault was greatly taken with using the term 'technology' to retrospectively describe the broad sweep of his work. This term is a little surprising given that Foucault uses it to name investigations which are primarily grounded in either the examination of discourse or an analysis of power. What Foucault sought to do was to bring these distinct foci within the same analytic frame by naming them as instances of a 'matrix of practical reason' (Foucault 2000: 223). For example, an investigation of the meaning of the notion of 'population' as it develops in the eighteenth/nineteenth century takes as its object a systematic set of procedures for knowing, describing and organizing a national collectivity. The emerging discourses of social welfare couched in terms of population are of a piece with the technical aspects, such as the development of social administrative practices, although neither is reducible to the other. The expansion of the term technology far beyond its usual referent of material equipment to capture the relationship between practical reason and the structured practice in which it is enacted is also present in the work of Heidegger (1977), a major influence on Foucault's so-called mid-period work (e.g. Foucault 1970).

---

3   Becker and a number of other senior managers at Humanitas have studied for PhD/ DBA's at the Universiteit voor Humanistiek, Utrecht where the second author (Brown) is visiting professor.

In recent years, Foucault's respecification of technology has been broadly adopted within Science and Technology Studies (STS), in particular within Actor-Network Theory (ANT) (see Latour 2005; Law 2002). From this perspective what we call 'the social' is in fact a series of technical relationships that mediate the interaction of actors (who are themselves a mixture of humans and non-humans). For example, in Annemarie Mol's (2002) studies of hospital environments, there is not a clear separation between doctors and patients. Rather there are shifting arrays of relations – xrays, tests, forms of knowledge, bodies, touches – which produce 'patients' and 'diseases' as provisional entities in an ongoing 'ontological choreography'. The patient – what and who they are, what conditions they may be said to exhibit, what the prognosis is for the future – is thereby produced within the technical relations of the clinic on a moment-to-moment basis. This implies that the body of the patient is never a single entity as such, but rather a series of possibilities which become actualised in various ways during the course of their progress through the clinical system. A less dramatic position, which is nevertheless consonant with a Foucauldian view of technology, is to see technical arrangements as concretised or congealed forms of social relations (as with Langdon Winner's 1980 celebrated discussion of the apparent social purposes of Robert Moses' urban architecture in New York). A bridge is then not simply a technological marvel, but also a way of reifying the relationships between the communities that it links. This is more or less the default view of technology adopted in contemporary STS.

If technology can already be said to be social in this sense then what is gained by adding the apparently redundant prefix in the term 'social technology'? The perspective adopted by Derksen and Vikkelsø is that there are some forms of technology which not only have a relatively immaterial basis in their enactment, but which also are explicitly, rather than implicitly, directed at transforming social relations. For example, the practice of psychotherapy is based primarily on the conversation of therapist with client (although it is of course mediated by the technical relations of couches, appointments and insurance arrangements). The goal of therapy is the transformation of the consciousness or emotions of the client. Hence we may define a social technology as *that which takes as its primary object the modification of some subjective state of affairs of a human subject*. Doubtless this is a wide and perhaps unwieldy definition, but it does allow us to retain another key notion for Foucault that technologies constitute rather than simply reflect the states of affairs they bring about. For example, if the aim of therapy is to manage self-loathing, then we may treat this phenomenon as itself shaped and articulated within the therapeutic process, rather than a fully formed object that is brought to the therapeutic encounter. Indeed we might say, following Foucault, that social technologies are the means of constituting subjects themselves, of 'making up people' in various ways, as Ian Hacking (1995) calls it.

From this perspective, we see the reminiscence museum as a social technology that is designed to modify the states of affairs of the visitors (and staff) who engage with it. In this case the states of affairs concerned are those which define the relationship to one's own past; in other words, memory. The museum is a technology for transforming what can be made of the past in present circumstances. It does this by affording a space where recollections can be articulated in conversation with others and via the mediation of the particular design of the space and the arrangement of the objects collected there. There is of course just such a practice which has existed for some time known as 'reminiscence therapy'.[4] The reminiscence museum at Akropolis was in part inspired this practice. However unlike reminiscence therapy as it is traditionally practiced, and in line with the art of living as Humanitas defines it, the museum does not have any therapeutic goal. Nor does seek to encourage remembering as means for the elderly to come to terms with the limitations of their present situation with respect to the totality of their life (the 'life review' aspect of reminiscence therapy). Finally, whilst reminiscence therapy is premised on surfacing personal recollections, the sorts of memories which may be described during visits to the reminiscence museum do not have to be considered as veridical accounts of the past.

This last point is drawn from a treatment of memory known as *Social Remembering* (see Middleton and Edwards 1990; Wertsch 2002; Middleton and Brown 2005; Reavey and Brown 2006). Drawing on sociological and anthropological understandings of memory in figures such as Maurice Halbwachs and Frederic Bartlett, social remembering emphasises that memory is reconstructive as well reproductive. The primary concern is not with the accuracy or not of some memory, but with how social practices of remembering invoke a version of the past to accomplish some social act in the present. For example, with regard to Karin's utterance 'home' in our opening example, the point is not whether or not the reminiscence museum triggered some memory of home, nor is it with a possible confusion between past and present. Instead what is most significant here is what it means for Karin to have invoked the notion of home and what that might achieve in the here and now as she encounters the museum space. It further follows that questions of what elderly people can remember and the impact that their advancing years might have on the clarity of the memories are also not of central concern. As we will see later on what matters is what is accomplished in the present in the course of a jointly managed act of remembering.

We will now turn to analyse some material drawn from an ethnographic study of the museum conducted by the first author. Material was gathered through participant observation as a member of staff based in the museum over eighteen months. It was supplemented by observation of a variety of other activities within Akropolis around the opening of the reminiscence museum, along with interviews with the director, the curator, staff and visiting clients. In the following section we

---

4 The idea of the museum emerged when Becker, inspired by reminiscence work practiced by some staff members, had an old Singer sewing machine and a bean cutter belonging to his mother brought into his office. Since his office has an open door policy, the objects quickly became of interest to a wide number of people throughout Akropolis, and appeared to elicit the sharing of memories and telling stories. The museum was envisaged as an extension of this process.

will look at some transcribed extracts taken from recordings of visits, where staff and clients talk together. After this we return to the question of how the museum might be said to support an 'art of living' in later life.

Mobilising the Past in the Present

The reminiscence museum is open every afternoon for six days a week. It may also be opened in the mornings by prior appointment to accommodate large groups. The museum is accessed through a lift near the main entrance in the Akropolis home. Beside this lift is a small glass display which contains domestic objects. These are changed throughout the course of the year to reflect the seasons. The display marks the transition between visual clutter of the 'village square' and the different kind of spatial arrangement at the museum. From the lift visitors enter via a large ornate wooden door into a main room that contains a wide variety of objects and seating. Typically visitors linger in this entrance room for a while, reacting to the objects there (which include a 1950s style moped). At this point visitors may approach or be approached by staff, depending on the number of visitors in the museum. Visitors progress through the museum slowly either accompanied or unaccompanied by staff, with visits usually lasting in the region of an hour. In the following extract an elderly woman visitor (W) is viewing the 'kitchen' space with a female member of staff (MW):

Extract 1

W: Once a week a big tub came into the kitchen.., the hot water went in, and then the three of us went in... The first one who went in, it was the same water. And there was always a fight who was allowed to go in first (*both laugh*).

MW: Wasn't it that the youngest always...

W: No, no, no.

MW: How many children were you?

W: Seven children. ...

W: I mean, it was a large zinc tub and that it, well not full of course but a couple of.. pans with hot water went in and you have to heat it on the gas stove. And that cooled off quickly, and now I must... can I have some more? (*Laughs*)

MW: Yes, yes, and that was just a bar of soap, wasn't it?

W: Or the shower gel, something like that.

MW: Yes, yes.

Vr: Eens in de week kwam een grote teil in de keuken.., kwam het heet water daarin, en dan gingen we met z'n drieën in...De eerste die dan ging, het was hetzelfde water. En het was vechten wie de eerste mocht (*lachen allebei*).

Mm: Was het niet zo dat de jongste altijd...

Vr: Nee, nee, nee.

Mm: Met hoeveel kinderen waren jullie?

Vr: Zeven kinderen....

Vr: Ik bedoel, het was een grote zinken teil en dat die, nou vol niet natuurlijk maar een paar..pannen heet water gingen er maar in en je moet het op het gas warm maken. En dat was ook snel koud, en nou ik moet.. kan er nog een beetje bij? (*Lacht*)

Mm: Ja, ja, en dat was gewoon een zeepblokje, he?

Vr: Of de douche gel, zo iets.

Mm: Ja, ja

W: And the washing glove, you have to wash … as well.

MW: But that wasn't every day?

W *almost whispering*: No, on Saturdays we went into the tub. Yes.

Vr: En de waslap, moet je ook …wassen.

Mm: Maar dat was niet iedere dag?

Vr *bijna fluisterend*: Nee, s'zaterdags gingen we de teil in. Ja.

The sight of a large tub in the kitchen area provokes an act of recollection on W's part. She formulates a small description of scene of collective bathing, where children share bathwater. The scene is nicely self-contained with a humorous description of the 'fight' which would ensue between children. This serves as an invitation for MW to join in the laughter, which she does and then builds upon the recollection by asking for further details. In her questioning MW also attempts to display 'being in the know' about such arrangement by asking whether the youngest child would be given the privilege of going first in the tub. The recollection is then jointly built by the two women, with W offering a further humorous vignette where a child protests at the lack of hot water. Following this some more details are requested by MW and provided by M before she offers a completion to the recollected episode, signalled by lowering her voice and returning to the opening statement that the scene occurred once a week.

What is significant about this episode is that W situates herself as a child within the recollection. Presumably it might as well have been possible to tell a story of her own adult experiences with such a tub (perhaps bathing her own children). W builds up the scene from the perspective and concerns of a child (e.g. fighting over turns in the tub) and uses active voicing to dramatise events further ('can I have some more?'). The domestic scene is then unfolded around the tub, which serves as the marker between then and now. Little description is offered of her family itself – instead it is the relations between persons as mediated by objects such as the tub, hot water and washing gloves which provides the focus. The tub present in the kitchen then acts as a kind of envelope (see Middleton and Brown, 2005) into which this description of a typical Saturday bathtime is packed. Once that typicality has been adequately described it returns back to the tub itself, and hence to present.

In extract 2 we see a different recollection which emerged around the same object. This time there are four elderly women visitors (N, N2, W3, W4) accompanied by a staff member (MW). Unlike extract 1, where MW played an essential role in inviting further details, here the women collectively build a narrative around the tub for the most between themselves:

Extract 2

N: Do you see that tub?

W3: Yes.

N: We used to have a very large one in the garden.

N2: And we have to go into the tub in the evenings.

N: Exactly.

N2: Not every day.

N: No.

All together

Mw: How many of you at the same time?

N: One at a time but on the same day.

Mw: Yes, of course.

N: Fetch water at the water boiler.

Mw: And how many children then?

N: We were with five children at home. ... We were divided. My sister was the last-born one, so she didn't need to, she went in one of those small tubs (Mw: Yes, yes, yes). And I had to go with my three brothers on Friday afternoons and with my two sisters on Wednesday afternoons.

Mw: Yes, that was well arranged.

N: ... fetch water first ... carry buckets with boiling water.

W4: At the water boiler's.

N: Yes, at the water boiler's . I lived in the staircase (W4: Yes). And I really walked up the stairs with the buckets with hot water. I was, I am taking about when I was 8–9 years old. (W4: Yes) With that boiling water...

Mw: That is the way it was, everybody did it.

N: I never burned myself.

Mw: No, no, indeed.

Vr B: Zie je die teil?

Vr 3: Ja.

Vr B: We hadden een heel grote in de tuin.

Vr B2: En we moeten 's avonds in de teil.

Vr B: Precies.

Vr B2: Niet elke dag.

Vr B: nee

Door elkaar

Mm: Met hoeveel tegelijk?

Vr B: Een tegelijk maar wel dezelfde dag.

Mm: Ja, natuurlijk.

Vr B: Water bij de waterstoker halen

Mm: En hoeveel kinderen dan?

Vr B: We waren met vijf kinderen thuis. ...We waren gedeeld. Mijn zusje was het nakomertje, dus ze hoefde niet, zij ging in zo'n klein teiltje. (Mm: Ja, ja, ja) En ik moest met mijn drie broers vrijdagmiddag en met twee zusjes op woensdagmiddag.

Mm: Ja, dat was goed geregeld.

Vr B: ... eerst water gaan halen... met emmers heet water sjouwen.

Vr4: Bij de waterstoker.

Vr B: ja, bij de waterstoker. Ik woonde in het trappenhuis. (Vr 4: Ja) En ik liep echt met de emmers heet water de trap op. Ik was, ik heb het toch over dat ik 8–9 jaar was. (Vr 4: Ja) Met dat heet water...

Mm: Het was niet anders, iedereen deed het.

Vr B: Ik heb me nooit van verbrand.

Mm: Nee, nee, precies.

This collective building of a narrative around the tub is noticeable from the opening of the extract. Whilst it is N who open up the topic (using a classic adjacency pair formulation of asking a question designed to move her to a substantive third turn – see Sacks, 1992), it is N2 who builds the context of the tub being a device for washing, which is then confirmed by N who goes on to take receipt of the questions addressed to the group by MW. N then offers a small description of bathtimes, again from the perspective a child rather than as an adult. As with the previous extract, this description focuses on practical arrangements that the child endures during bathtime, such as carrying buckets of boiling water.

The contribution by W4 is significant as she offer the phrase *waterstoker*. This is local Rotterdam dialect for a water boiler. Hence the receipt 'ja, bij de waterstoker' by N in the next turn does membership work. The recollection is placed as relative to a common past that is shared by the women of growing up in the local area where the museum is now situated. Jointly building the recollection then does a work of constituting this group of women in the present as sharing community and a communal history.

But interleaved within this building of community there is also a significant personal contribution made by N. MW – the staff member – offers a reflection (and appreciation) of the communal bathing at the end of N's first long turn ('Yes, that was well arranged'). N could have responded by moving on from the child's perspective, or expanding out to a broader description of domestic life. Instead of this, she responds by detailing further the child's part in this 'well arranged' scene. It becomes apparent that as a child N found this to be anything but a satisfactory arrangement. She emphasises potential dangers – 'carrying buckets with boiling water' – along with the extraordinariness, from the perspective of the present, of the ordinary arrangements of sending a child ('I am talking about when I was 8–9 years old') to climb stairs with scalding hot water. MW's subsequent turn ('That is the way it was, everyone did it') does a work of normalising such a practice historically, thereby both affirming N's story and acknowledging its extraordinary nature in terms of contemporary views of childhood and adequate parenting.

We might see N's story as a kind of complaint, as a critical reflection on the past. Yet if we consider its place as the part of the interactions occurring during the visit to the museum a different sort of interpretation is also plausible. Note the conclusion to N's story – 'I never burned myself'. This completion renders the story that has gone before not as a complaint but as evidence of personal autonomy in the midst of vulnerability. As a child, N experienced bathtimes as dangerous and arduous, but she accomplished the tasks she was obliged to do without ever coming to harm. The story is told by an elderly woman who is now herself vulnerable. The hearable implication of this story is that the strength and autonomy of the child remain in the elderly woman she had become. And insofar as the story is woven into a collective shared past that attribute may be taken as representative of the group as a whole in their shared present moment.

This doubling of past and present around the specific episode of the *waterstoker* was also found in the following extract:

Extract 3

W: And then she (*granny*) was living on the third floor, and then she went to the... water boiler's , she went to fetch buckets.., such buckets with hot water, she would climb up all those stairs with the boiling hot water. It went into that and then it started to ... I have done all that too as a child, where you get the water, yes.

Vr: En dan woonde zij (*oma*) driehoog, en dan ging ze bij de... waterstoker, ging ze emmers.., zulke emmers met het hete water halen, liep ze al die trappen op met het gloeiend hete water. Dat ging daarin en dan ging het een beetje.. ik heb het ook allemaal gedaan als kind, waar je het water haalt, ja.

Here the switch between age and youth is done explicitly. W begins her description by invoking her own grandmother, but then rapidly switches to a description of climbing the stairs as a child. Thus in very quick succession W makes links between her own present age to that of a then elderly relative in her childhood who was nevertheless still able to carry the hot water bucket, through to her own childhood experience of doing the same. Personal autonomy is doubly emphasised in the past and thereby securely made relevant in the present.

Switches in age and perspective as recollected episodes unfold are also apparent in the next extract. This is taken from the transcription of a group visit to the museum. Five women (W1–5) were accompanied by two professionals (MW) from another nursing home. This small conversation takes place in 'the kitchen', where one of the women sees an old-fashioned moneybox:

Extract 4

W3: ...That thing standing over there, my mother had that.

W2: Yes, I had that too at the time, that was from the savings bank.

W3: Was for the gas and light bill.

W2: Ye-es, had that myself.

W3: .. and for clothing.

W2: Yes.

W3: We didn't do about holidays but the delivery boys came to the door (W2: Yes). I remember the landlord (*all together*)..., undertaker...

W2 *simultaneously*: I began at ... 20, or something like that, can remember it well.

Talk all together. Pause

W2: How life has changed, hasn't it! When you see this then. On Mondays my mother never left the house. Everything came by there.

Vr3: ...Dat ding die daar staat, die had mijn moeder.

Vr2: Ja, dat heb ik ook nog gehad toen, dat was van de spaarbank.

Vr3: Was voor de post van het gas en licht.

Vr2: Ja-a, heb ik zelf ook nog gehad

Vr3: .. en voor de kleding.

Vr2: Ja.

Vr3: Vakantie deden we niet aan maar de bodes kwamen aan de deur (Mw: Ja). Ik herinner me de huisbaas (*door elkaar*)... begrafenis ondernemer...

Vr2 *tegelijk*: Ik ben begonnen me ... 20, of zo iets, kan me goed herinneren.

Door elkaar. Pauze

Vr2: Wat is het leven toch veranderd eigenlijk, he! Als je dan dit ziet. Op maandag ging mijn moeder nooit weg. Daar kwam alles langs.

W4: I can't remember any more on which day they came by.

W2 *almost simult.*: At our place on Mondays. Then everything came by.

W3: That used to come to the door.

W2: And when there were no cents left then she wasn't at home.

Laughs

W3: Then we still had to...

W2: Then we had a problem.

One of the volunteers goes upstairs with another client

W3: Hasn't everything changed, eh?

Mw: Now indeed, ho, incomparable. No.

W3: No, it is not comparable any more.

Vr4: Ik weet niet meer op welke dag ze langs kwamen.

Vr2 *bijna tegelijk*: Bij ons op maandag. Dan kwam alles langs.

Vr4: Dat kwam aan de deur vroeger.

Vr2: En als de centen op waren dan was ze niet thuis.

Lacht

Vr4: Dan moesten we toch...

Vr2: Dat was bij ons dan een probleem.

Een van de begeleiders gaat naar boven met een andere cliënt ...

Vr3: Wat is de heleboel toch veranderd, he?

Mm: Nu wel, ho, niet te vergelijken. Nee.

Vr3: Nee, het is niet meer te vergelijken.

The box in question is of a typical rectangular design used at the time, with separate compartments inside. The cover had a number of slots to insert coins. Very often the purpose of the savings was printed on top of the box, for example, coal, gas, clothing, holidays etc. Such a box was a standard accessory in many Dutch households. As such, it is one of the most frequently recognised objects in the Museum. In the opening turns W3 and W2 mutually build a description of the money box, pointing out its likely origin ('from a savings bank') and its purpose. In doing so they, like the women in extract 2, constitute a common shared past. But this common past is more complex than it first appears. For W3 it is the past she remembers as a child, with the moneybox being in the possession of her mother. For W2, by contrast, it is the past she recalls as an adult when she herself used a similar moneybox. In one sense this is not the same historically defined portion of the past at all. And yet it is the same 'past in general' which is built and jointly experienced from two very different perspectives without ever disturbing the commonly built narrative.

This switching between perspectives continues in the following turns, until W2 herself shifts a generation to discussing her own mother. This builds to the climax of the story when the procession of tradesman visiting the house has exhausted the reserves in the moneybox. At this point the children answering the door are required to inform visitors on their mother's behalf that 'she wasn't at home'. The switch made by W2 may then be strategic – the joke is not funny when told from the perspective of the mother, where it takes on the appearance of complaint or tragedy. But what then has 'changed' or been rendered 'incomparable' between then and now, as W3 proposes in the final line? This could be taken as a remark about financial security – a precarious past represented by the dwindling resources in the money box has been supplanted by a secure present. Or it could mean that it is now inconceivable that children should be made to solve the 'problem' created by an adult. Or yet again this might simply mark the generational transitions which are in play here, from child to mother to grandmother, all of which are freighted with great present significance. Whichever interpretation is followed all share the common feature that a version of the past is here being mobilised to establish direct contrasts with the present. The simple money box opens up onto a common past that can be told from a variety of perspectives and rendered as having a wide range of implications for one's present identity.

The final extract offers a different instance of how perspectives can be interleaved in the mobilisation of the past. The interaction transcribed here occurs in a particular part of the kitchen area which is filled with devices for washing and ironing clothes. This space is somewhat enclosed in comparison to the open kitchen area. An elderly woman (W2) and her middle-aged adult daughter (W1) peer into this space accompanied by a female staff member (MW):

Extract 5

In the washing corner, looking at the irons.

W1: These are 50 years old, because then you had them with... when Nicolette was born, then I had the first electric one.

MW: You are saying school. Was it an ordinary school?

W2: Housekeeping school....For girls it was a very good education in those days, you know.

MW: Yes, yes.

W2: And the difference between Ingrid and my education is like between day and night. We were already sewing a coat when after two years... laying the table and serving, very tidy all that.

W2: Such a washing machine, had one as well. ...Yes, I used to have this one but I also have a wooden... And that one I had in the beginning of my married (Mm: Yes). I am 75 years, my daughters are 51 and 50, eldest.

W2: Oh yes, a big washing pan like this I put on the stove with water...and the washing was boiled in that too.

MW: Yes, yes, exactly.

W2: And then my husband would throw it in and it would stand on the stove soaking during the night and then in the mornings before he left for his work I would fill a wooden washing machine, but the spinning went a little bit different from this one, that flywheel was underneath

In het washokje, kijkt naar de strijkijzers.

Vr1: Deze zijn 50 jaar oud, want toen had je met..toen Nicolette geboren werd dan had ik de eerste elektrische.

Mm: U zegt school. Was het een gewone school?

Vr2: Huishoudschool. ...Toen voor de meisjes het was een heel goede opleiding, hoor.

Mm: Ja, ja.

Vr2: En het verschil van Ingrid en mijn opleiding is dag en nacht verschil. We naaiden al een jas als we twee jaar.., dekken en dienen, heel netjes alles.

Vr2: Zo'n wasmachine, ik ook nog een gehad. ...Ja, deze heb ik gehad maar ik heb ook een houten.. En dat had ik in het begin van mij trouw (Mm: Ja). Ik ben 75 jaar, mijn dochters zijn 51 en 50, oudste.

Vr2: O ja, zo'n grote wasbus ik heb met water op de kachel gezet...en daar werd de was ook in gekookt.

Mm: Ja, ja, precies.

Vr2: En dan gooide mijn man het daarin en stond het op de kachel 's nachts te trekken en dan 's morgens voordat hij te werk ging eerst dan laadde ik een houten wasmachine, maar het snel draaien deed ietsje anders dan deze, dat vliegwiel zat daaronder (she is speaking very fast now, almost without breathing) highly dangerous because.. my boy, unfortunately he died when he was 40, 4 years ago, but he was (very) as quick as anything and before you knew it he was fiddling at that flywheel.

    MW: O my God!

    W2: That's why, highly dangerous.

W2: My son who was 40, he would be 44 now, till he was 4, only then did I get a spinner. (MW: Yes, yes). But my daughter recently said: I still remember that you only ... they said then, fantasising: it is practically dry. (MW: Yes, yes.). It wasn't though. It wasn't that dry. (MW: No, no) But later on, my son is, my youngest son is 40, and then I says: I don't mind having a baby but the horrible thing was always to dry things in my house, I find it awf(ul), and then I got a dryer.

(praat snel, soort van zonder ademhaling) levensgevaarlijk want... mijn jongen, die is helaas overleden toen hij 40 was, 4 jaar terug, maar die was watervlug en hij zat al gauw even aan dat vliegwiel.

    Mm: O God!

    Vr2: Dus levensgevaarlijk.

Vr2: Mijn zoon die 40 was, die zou nu 44 zijn, tot zijn 4de jaar, toen pas kreeg ik een centrifuge.(Mm: Ja, ja) Maar mijn dochter zei laatst: ik weet nog dat jullie maar ... ze zeiden dan, fantaseerden ze: het is praktisch droog. (Mm: Ja, ja) Dat was toch niet. Dat was niet zo droog. (mm: nee, nee) Maar later, mijn zoon is, mijn jongste zoon is 40, en toen zeg ik: dat ik een kind krijg dat vind ik niet erg maar dat ellendige was altijd in mijn huis te drogen, ik vind het verschrik(kelijk), en toen kreeg ik een droogtrommel.

As a whole, the extract is driven the descriptions of the washing machines and devices given by W2. These list daily care routines performed for her children and her home. Each piece of equipment is indexed to a particular moment in her marriage and the birth and growth of her children. Early in the extract this role as homemaker is described with pride – her daughter may have had a different education but 'housekeeping school' was a 'good education in those days'. It prepared her well for the life she was to lead. However, midway through a description of washing her husband's clothes, a different tone enters into the recollection. A hitherto unmentioned third child – her youngest, a son – is mentioned. His introduction is doubly dramatic. His recent death is interleaved with a story about his child misadventures playing with the dangerous spinner on the washing machine. Her anxiety as a mother then ('before you knew it he was fiddling at that flywheel') is juxtaposed to the repeated details of his death. This is hearable in MW's shocked uptake – 'Oh my God!'. The concluding turn by W2 continues this alternation between the recent past and the more distant past, arriving at a strange resolution. It was not the ever present danger of the son injuring himself that was problematic about the old washing machine, it was in fact the 'awful' experience of having to dry clothes in the house. In arriving at this resolution W2 in effect tames the potential vulnerability that is invoked by the recent death of her son. One does what one can, but life is dangerous. It is not the danger to which I object, it is the ongoing inconveniences which we must endure...

This extract shares several features with the previous four. Scenes from the past are opened up in relation to a particular object. This then serves as the reference point or envelope around which family social relations are unfolded. What is invoked is not a highly personalised or unique past, but rather a past in general that is either jointly constituted by the speakers or is designed to be heard as commonly shared ('everything has changed', 'in those days'). The common past is nevertheless viewed from several perspectives, and through the experiences of different generations. Past and present are interleaved to either constitute some attribute that is continuous over time – such as personal autonomy – or contrasted between then and now – such as financial insecurity. Finally present circumstances and experiences are notable by the apparent absence in the conversation. One does not hear complaints about health or loss. But this does not mean that they are not there as ongoing concerns. Rather present dilemmas regarding autonomy, security, vulnerability and generational transitions are all thematised within the recollections themselves. Each episode is replete with possible implications for the present, and is indeed really only properly explicable from the perspective of being told by an elderly person, even if these implication are not spelled out directly during the interactions.

The Art of Living Revisited

We have described the reminiscence museum as a form of social technology. As such we have identified its general aim as being the modification of the subjective states of affairs of those who engage with it. This is accomplished through organising a very particular kind of space designed to facilitate joint and collective remembering. That work of remembering is not confined to reproducing or more-or-less accurately recalling the past. It is instead a complex and subtle reconstruction of the past for present purposes that is composed of multiple perspectives, generational shifts and contrasts along with juxtaposed themes and relations where present concerns are worked by selectively invoking versions of past episodes and scenes.

The question then remains the extent to which all of this can be considered to comprise care, in the traditional sense of that term, or if it is instead best viewed from the 'art of living' defined by Becker and Humanitas. First of all, we hope to have shown that the reminiscence museum differs markedly from the practice of reminiscence therapy. The museum espouses no therapeutic goals, nor is it equipped in any way to support them. There is no attempt to guide visitors in their recollection, nor is there encouragement for them to view the past as a coherent story which defines their present circumstances. Outcomes are neither defined nor monitored[5]. Visitors' engagement with the space is not prescribed beyond that of common courtesy. This is not the provision of care in the typical sense of 'cure' and 'care'.

---

5   There is one exception here. Memories of the destruction of large parts of Rotterdam during the second world war and the subsequent 'hunger winter' remain strong in the local community. In the rare cases where visitors to the museum become obviously distressed in the course of recalling these events there are several residents who are known to be skilled at discussing and diffusing such memories who may be summoned to intervene.

Nevertheless, the vast majority of visitors to the museum have an overwhelmingly positive experience, and many do indeed feel 'cared for' either by staff, relatives, or perhaps by the space itself during the course of their visit. The following description of a visit is based on field notes made by the first author:

> On the 22nd of November 2006, a couple of months after my project had started, I went to one of the departments at the Humanitas nursing home and asked whether I could take a patient with me to visit the Museum. The Museum was not officially opened yet, but within the yes-culture of Humanitas that kind of suggestion is never a problem. The people who were living in that department were suffering from both physical and dementia problems. The department nurse looked around and then suggested that I could perhaps try and take Mr Sharp with me. Mr Sharp was 79 years old; he was sitting in wheel-chair and was suffering from aphasia. I introduced myself and asked him whether he would mind going to the Museum with me. He kind of nodded, I took it for a *yes* and we went downstairs. We started the round in the grocery shop. I asked him some questions and he did his best to remember things and to provide the right answers. His replies were never sentences, mostly one-worded, but the words were the 'right' ones. Initially (at least this is what I thought then) his facial expression did not change very much. I made sure that he could see and hear me in front of him, while showing things and taking the time (too little as I understand now) before he could work out his answer. We were clearly two separate islands: one of us curious and investigative, actively gathering live material for a thesis, while the other was isolated within himself, vulnerable and simply complying with the intrusion. We moved to the sitting room and he suddenly showed a clear interest in the music that was playing on the old-fashioned radio. He was almost literally struck by the sound of recognition. From my side I was struck by the way the Museum could work. In the kitchen I put an old-fashioned coffee grinder into his hands. He did his best to turn the handle, but it was clearly too heavy for him. 'Too heavy', I registered… . And then he reached out to me and touched my hand.

> Mr Sharp could not do very much. His touch was soft and non-intrusive. He could not hold his hand steadily, but moved it softly, which gave me the impression that he was stroking my hand. He smiled in a friendly and inviting manner, encouraging in fact. I did not realize at that moment what had happened but all at once everything had changed. Today I would say that it was he who awakened me that day, and not the reverse. I started looking at his face twice as often, in order to follow his gaze and anticipate what he was recognising. He was not silent, as I had been thinking only moments ago. His eyes, his expression and the subtle motion of his lips, everything in his posture was talking to me all the time. Only I had not seen that at first, since I had been busy with my personal agenda. I followed his gaze and pushed his chair to an old bakelite telephone. When I put the heavy receiver in his hands he was clearly delighted. He touched the sturdy

object and I saw this stroking motion again. We moved on and his look stopped me at the school corner. I took one of the old copybooks and anticipating his desire I placed it into his hands. He held on to the copybook tightly, constantly touching and stroking the old thing, with a shadow of a smile lightening his face. It took quite some time before I felt that we could move on again. The next stop he wished to make, which he made clear to me somehow, was in front of the wardrobe in the bedroom. Once again I followed his gaze and took a stiff collar from the shelf. That time he not only tried to show how you closed it, but also reached in the direction of his neck, where he wanted it to be placed. He did the showing, while I did the talking. We were a team. We were conversing.

Then it was time for us to go back again. We were waiting together in front of the lift when he took my hand again, brought it to his face and let it lie on his cheek for some time. By softly stroking my hand he was saying thank you, and I, despite being armed with all the communicative power of language, did not know better than to stroke him back and to hold him tight.

During the course of the visit, various kinds of small transformations occur. Mr Sharp moves from apparent disinterest to engagement with the objects. The researcher shifts from viewing Mr Sharp as debilitated and uncommunicative to active and purposive in his exploration of the space. What the space is, what it can be, and what it means to experience it as a client or as a researcher also changes. Here we can see two of the threads that Becker captures with the 'art of living' – social involvement and a critique of existing representations of ageing. Both are performed and become intertwined as the visit progresses. As a consequence, the third thread – giving meaning to existence – certainly seems to be accomplished, even if in a purely provisional, temporary way. Finally the closing scene in the list certainly speaks to Becker's more open aim of 'making people happy'.

Yet we want to venture that something else is happening both in this particular visit with Mr Sharp and within all of the others discussed above. If a social technology acts to construct subjective states of affairs as well as modifying them, then we should not see the 'art of living' as involving a transformation from one given pre-defined state (i.e. passive) to another (i.e. active). It is rather the case that the museum may problematise and redefine the nature of the visitor. Mr Sharp is not suddenly a great non-verbal communicator as a consequence of his visit. Instead the visit provides the opportunity for a relative redefinition of the present through the mobilisation of the past. As he strokes the copybook, the apparent trajectory from past activity to present passivity is disrupted, it becomes more difficult to sustain. We might say then that the art of living consists not of transforming the present into a new future, but rather of opening up the past from its apparent rigidity and determination to afford new possibilities in the present.

This reading of the art of living, or as it was once called *tekhnē tou biou*, is close in spirit to that given by Foucault in his final works. There he speaks of how structured practices for working on oneself, for producing oneself as a particular kind of subject evolved in ancient Greece and Rome. The goal of this 'art of existence' was to 'permit individuals to effect by their own means, or with the help of others, a certain number of operations on their own bodies and souls, thoughts, conduct, and way of being, so as to transform themselves in order to attain a certain state of happiness, wisdom, perfection or immortality' (Foucault, 2000: 225). We moderns hope for less, of course, but perhaps share in the ambition to be able to participate in a collective work which may result in a relative freeing up of our 'souls, thoughts, conduct' through the mediation of social technology. We do not expect immortality, nor to avoid the inevitable ravages of ageing. But we can hope for our lives to not be written for us in advance, for there to be no assumption that our present is defined by a past that is fixed, immutable and holds us to what we apparently are. Insofar as the reminiscence museum affords the opportunity to do this, then it is indeed in the service of an art of living in both Becker and Foucault's sense of the term. It proposes that the operation the elderly stand most in need of is not biomedical, it is instead that performed on their very way of being.

References

Becker, H. 2003. *The Art of Living in Old Age: Happiness-promoting Care in an Ageing World*. Delft: Eburon.

Foucault, M. 2000. *Technologies of the Self. In Essential works of Foucault 1954–1984*, Vol. 1. London: Harmondsworth: Penguin.

Heidegger, M. 1977. *The Question Concerning Technology. In The Question Concerning Technology and other essays*, trans. W. Lovitt. New York: Harper and Row.

Latour, B. 2005. *Reassembling the Social: An Introduction to Actor-Network Theory*. Oxford: Oxford University Press.

Law, J. 2002. *Aircraft Stories*. Durham: Duke University Press.

Letiche, H. 2008. *Making Healthcare Care: Managing Via Simple Guiding Principles*. Charlotte, MC: Information Age Publishing.

Middleton, D. and Brown, S. D. 2005. *The Social Psychology of Experience: Studies in Remembering and Forgetting*. Sage: London.

Mol, A. 2002. *The Body Multiple*. Durham: Duke University Press.

Reavey, P. and Brown, S. D. 2006. Transforming Agency and Action in the Past, into Present Time: Adult Memories and Child Sexual Abuse. *Theory and Psychology*, 16(2), 179–202.

Winner, L. 1980. Do Artifacts Have Politics? *Daedalus*, 109(1).

Chapter 9

A Pillow Squirrel and its Habitat: Patients, a Syndrome, and their Dwelling(s)

Research Center for Shared Incompetence / Xperiment!

B. Kraeftner, J. Kroell, G. Ramsebner, L. Peschta, I. Warner

Pillows – an Introduction

In many cases we consider a pillow an important part of our very private and personal world. Maybe a pillow is a kind of transitional object. It helps to make the transition or the relationship between me and the world more pleasant and cushy, softer and smoother: more *pillowy*. Irrespective of whether we think of the obdurate world of objects that we stumble across by leaving the protective envelopment of the uterus or by entering our sleeping berths; or whether we think of the elusive world of thoughts and dreams that we rarely visit without the company of a pillow – or several pillows – which may themselves become, now and then, important and active parts of those worlds. In a sense, one could argue that pillows are part of our bodies and souls.

Yes, of course, we allude to D.W. Winnicott's concept of ordering the relation between inside and outside, objective and subjective realities (Winnicott 2002). Aside from pillows, we call to attention those artefacts that we all know: dolls, teddy bears, soft toys of all kinds - blankets. And we call to attention the emotional domain that is frequently and inextricably associated with those transitional objects and that takes part in those modes of ordering. This allusion, for us, is not an explanation but an opportunity to sensitize ourselves to the intricate relationships between humans and artefacts.

Pillows are physical-technical artefacts. We refer to auxiliary material in a gravitational world that, in various health care settings, helps to keep patients in varying positions with the aim of preventing the development of pressure sores, to provide a means to relax the limbs and body, and to establish the possibility that patients can adjust their position in relation to the world. (Figure 9.1)

You may have noticed that we have entered the realm of care and technology. Accordingly, we neatly separated the description of pillows into two domains: first, pillows and private care for ourselves and/or our caring relationships to others at home; second, pillows as part of a technological framework and instrumental relations at the workplace. This separation does not reflect our inclination to isolate care from technology or vice versa. However, it gives us the opportunity to denote the dichotomous structure within which discussions of care and technology frequently unfold.

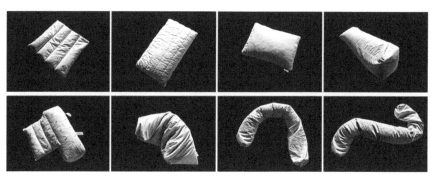

Figure 9.1   Pillows for patients

In the following chapter we – members of a transdisciplinary research group[1] – describe an ongoing research project that we call pillow research and that follows the opposite direction: it welcomes technological artefacts and the irreducibility of the intricate relationships between care and technology and its decisive role of enacting the care for a syndrome (Kräftner et al. 2010). It describes our attempt to introduce artefacts ("pillows", "transitional objects") into the world of long-term care of severely disabled people. The (artistic) tinkering with those artefacts within the clinical context should experimentally create conditions that contribute to the emergence of unexpected spaces: spaces of opportunities to enrich the dwellings of patients, moments of unforeseen relations between private and institutional spaces, new spaces of (in)dividualities (Strathern 1992).

There, where multiple tensions dominate the practices of everyday life on a specialized ward, one faces questions like this: Which syndrome do "we" want – here the "we" designates members of the multidisciplinary team, members of the patient families or of a research discipline (since we face not a pre-given syndrome but more or less dominant versions "in the making" – see later)? Who takes part in the diagnosis, treatment and care of patients? What kind of research – if any – do we prefer and how does it enact bodies, subjectivities of patients and caregivers? How should we handle errors and failures and the continuation of *our* work? How do we rank the importance of the development and training of skills, gaining hands-on experience and the constant modifications of minute details of

---

1   This account of events should give an impression of our involvement as transdisciplinary artists-researchers interested in the exploration and combination of scientific and artistic methods. The first phase of the project started with an (visual) ethnographic research approach called "Performing Shared Incompetence – A Topography of the Possible. What is a Body / a Person?" conducted by the research group Xperiment! / Research Center for Shared Incompetence (RCSI) within the transdisciplinary research program 'TRAFO' of the Austrian Ministry of Science and it was displayed in various exhibitions (see Latour 2006; Xperiment 2005). The current phase has developed as "Pillow Research" financed by the Austrian Science Fund (FWF). For the publication of the images, informed consent has been secured.

care? Of course, this chapter does not answer these questions; in fact, it describes a part of the pillow research group's work and its tinkering with one exemplary pillow – or transitional object – that should give us the opportunity to elucidate our methodological approach.

The remainder of this chapter is organized in five parts: first, we describe various assistive and conventional (care) technologies that can be encountered in the respective care context and that form part of the dwelling of these patients; second, we discuss several technically heterogeneous diagnostic techniques that quite recently started to modify the definitions of the syndrome; third, we discuss the different logics of care that may come along with the implementation of different technological approaches; fourth, we would like to prepare the reader for our exemplary case study by making a historical detour and by offering a risky bet that will lead us back to, fifth, the introduction of a technical pillow design and a distributed, situated and collective diagnostic setting that explores the meaning of "diagnosing/assessing" that contribute to the discussions of the aforementioned questions.

Bodies/Persons and their Dwelling(s)

People affected by this syndrome (which – by the way – has no diagnostic key in the ICD-10) do not talk or move and apparently are unconscious for months and sometimes for years. They are legally incompetent persons/bodies who are cared for at specialized wards, ordinary nursing homes or – rather unusually – at home. We are speaking of a syndrome that is called vegetative state. This term refers to persons who have undergone severe brain injury and subsequently remain in a state of prolonged coma. After some time, they open their eyes and appear to be awake and furthermore, show preserved sleep /wake cycles. The definition says that those patients do not show any reactions to their environment or themselves. They are considered to be devoid of consciousness and awareness. (Jennett 2002, Andrews 1996) In the following we refer to practices that can be observed within an institution that hosts approximately 36 patients and that specialized in the alignment of various versions of the syndrome to form a single "vegetative state".

The appearance of a body/person and its structure may depend on the surroundings and environment it occupies. The specific patients who suffer from vegetative state almost exclusively have two square meters each at their disposal. Even nowadays it is not uncommon that these patients almost never leave these dwellings more commonly referred to as "bed". (Figure 9.2) Two square meters – 24 hours, day after day, lasting for weeks, sometimes for years on end. It seems self evident that this fact affects form and function of a body/person. In textbooks and monographs, written forty years ago, we find photographs of patients where we can examine the results of this „lifestyle". In general, these images show a cachectic and distorted body; and, furthermore, one can recognize a recurrent

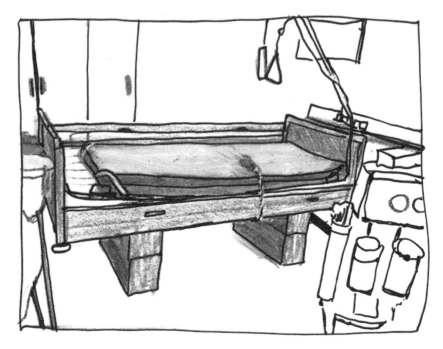

Figure 9.2   Casa Comfort Plus®

phenomenon that is caused by the body's/person's inability to leave its home: i.e. the destruction of its integument due to constant pressure of the body's own weight on the flat base of its dwelling. (Gerstenbrand 1967) During the last decades several technological measures were deployed to take care of these bodies.

Assistive Care Technologies

Regarding the syndrome one could describe these measures as never-ending experiments in changing the surroundings and the respective dwelling places to help in "reshaping" these bodies. First, there are attempts to modify the place itself: many technological designs aim to improve the quality of accommodation and reducing the risk of pressure sores. For instance, some bodies are kept floating on air to keep their integument in good order. However, the beneficial effect on the skin has to be weighed against the problem of the patient's potential loss of any sensation of her or his own body – a loss that may cause or aggravate stereotypical motor behaviors. (Steinbach and Donis 2004)

Second, there are increasing efforts to help patients to leave their dwelling places since caregivers consider it important to avoid sensory deprivation. Therefore patients are transferred to (wheel)chairs and brought to different places inside or outside the ward. The very act of transferring the patient can be accomplished by means of human or non-human lifting technologies. A so-called non-human technology is an electro-mechanical lifting device that makes the patient hover from bed to chair. This makes the life easier for nurses but is considered a stressful event for patients. (Schillmeier and Heinlein 2009) This is the reason why kinesthetic technologies have been recently developed to allow the caregiver's own body to become, together with the patient's body, a kind of living lifting device. However, this procedure runs the risk of breaking some osteoporotic bones of patients.

Third, during the last decade an important extension of the body was introduced to this field of long-term care: it helped to overcome the notorious difficulties in swallowing and to alleviate the problem of eating and food intake by simultaneously reducing the risk of pulmonary infections caused by aspiration. It involves placing a tube into the stomach through the abdominal wall so that nutritional support can be maintained. Tube feeding decisively "reshaped" the bodies and gave them the opportunity to gain weight and resistance to diseases. However, if one considers eating as an act that has something to do with tasting, smelling, feeling, interacting (with), or enjoying food – then tube feeding is a poor alternative for feeding or eating.

Fourth, and in addition to the alteration of the material-technical surroundings of the bodies, the living surroundings were re-organized: the number of persons who take care for these patients was increased. At some dwelling places the ratio between bodies and caregivers was increased to approximately 1:1.

In comparison to the bodies on the pictures taken forty years ago, today we can recognize remarkable differences in the appearance of some of the bodies: we see well nourished, much more relaxed versions of bodies with less severe distortions of the upper or lower limbs. The described differences are the result of efforts and experiences made during the last decade with the aim to find out where and how to accommodate those bodies/persons.

Of course, there are additional technologies that are experimental, as there are electrical bicycle-like devices that passively move the legs of patients; devices that bring and keep patients in an upright position; drug delivery systems using computer controlled pump administration; tracheostomy tubes to facilitate regular suctioning of the upper respiratory tract, etc. What is important to state is that almost none of these care practices, be they more or less technology driven, remain uncontested but rather give rise to debates.

Care(technologies) for the Syndrome

Maybe one reason for the debates concerning the care of apparently vegetative patients has to do with the fact that during the last years the answer to the question whether a patient suffers from the syndrome – according to its definition – has turned out to be more complex than assumed. Various (care) technologies have refined the "resolution" of diagnostic methods thereby delineating (we could say "enacting" (Varela and Dupuy 1992; Mol 2002)) various versions (Despret

1999) of the syndrome that until now were bundled under the term (permanent) vegetative state.

In the following we would like to describe various versions that come along with technological developments that converted these severely disabled patients into objects of interest for scientific research – patients who a few years ago "vegetated" without any notice in various nursing institutions. These versions address the (dis)interest in the absence or presence of consciousness and it will be interesting to find out whether and how the enactment of (dis)interest will contribute to a "reshaping" not only of bodies/persons but will also generate new spaces for renewed relationships and interactions between patients and care givers.

Care for an Embodied Version of the Syndrome

In the Western countries persons in vegetative state are rarely cared for at home. At specialized health care institutions mandatory 24–hour care is provided predominantly by registered nurses. In German-speaking countries this nursing understands itself as therapeutic nursing, beyond the mere fulfillment of basic needs like feeding, cleaning and keeping warm. We are speaking of nursing practices which include therapeutic techniques and interactions that follow "functional exercising concepts" (Zieger 2003) like the "Affolter"- and the "Bobath"-concept", "Basal Stimulation" and "Kinaesthetics." These rehabilitative techniques conceptualize the living body as a whole that perceives and acts beyond cognitive capabilities. In other words, the question whether the patient shows any behavior that can be associated with conscious behavior remains in the background. For these caregivers patients may express their mood, will, refusal, affection, decision-making, etc. by behaviors like being "tensed up/ relaxed", "congested/non-congested with phlegm", "calm/anxious", "out-of-tune/ happy", "sleepy/awake"; even the signals of the autonomic nervous system like body temperature, sweating, breathing (frequency), heart rate, bowel movement can be interpreted as communicative signs. These apparently vague and fuzzy categories are important reference points, in order to guide any nursing behavior. In addition, since in German-speaking countries withdrawing fluid and nutrition is not an option, the fact of whether those patients are "devoid of consciousness and awareness" remains without consequence.

Care for a (self-)Conscious Version of the Syndrome

In other countries the question of whether a patient is in a vegetative state or not, whether she or he shows signs of awareness or not, is of great importance since there exists the option of withdrawing fluid and nutrition. As soon as any signs of a patient's awareness emerge, this option ceases to exist. This issue was raised when, during the nineties, a study in the United Kingdom showed that:

The vegetative state needs considerable skill to diagnose, requiring assessment over a period of time; diagnosis cannot be made, even by the most experienced clinician, from a bedside assessment. Accurate diagnosis is possible but requires the skills of a multidisciplinary team experienced in the management of people with complex disabilities. Recognition of awareness is essential if an optimal quality of life is to be achieved and to avoid inappropriate approaches to the courts for a declaration for withdrawal of tube feeding. (Andrews et al. 1996: 13)

These concluding remarks of the study referred to the surprising finding that up to 43 per cent of the patients were misdiagnosed as being in vegetative state. Two statements foreshadow a current diagnostic trend that co-evolves with the development of new technologies: first, "diagnosis cannot be made, even by the most experienced clinician, from a bedside assessment:" this statement refers to the deplorable fact that in many cases a single behavioral assessment conducted by a physician suffice(d) to return the verdict: "permanent" vegetative state. It further suggests that time, training and experience is a prerequisite to become familiar with the patient. Second, "recognition of awareness is essential:" this statement shows that any hint at the presence of awareness and consciousness obviates a patient to be considered as being "merely" vegetative – "void of" any thoughts, sensations, percepts, dreams, sentiments, etc.

One can read these statements as a framework for two parallel research approaches that share a common goal: i.e. how to detect consciousness in people in vegetative state. However, these approaches pursue this goal through different technical means.

First approach: clinical neuro-cognitive assessments   The development of this technology aims at detecting consciousness by relying on a human observer who looks at the patient to discriminate any subtle signs of the recovery of consciousness. Of course, in contrast to a brief (neurological) examination the increased attention to any subtle but potentially important clinical signs lengthens the administration time. These assessment technologies are comprised of observations and the completion of standardized forms; these comprehensive checklists order any observable behaviors in various hierarchical ways – an ordering that should guarantee inter-rater agreement and test-retest reliability. Furthermore, depending on the respective assessment techniques, various subscales and scoring levels should allow for differentiating patients in vegetative state from those in a so-called minimally conscious state (Giacino et al. 2002). Techniques are: the *Coma Recovery Scale-Revised* (GRS-R) (Giacino et al. 2004), the *Wessex Head Injury Matrix* (WHIM) (Shiel et al. 2000), *Sensory Modality Assessment and Rehabilitation Technique* (SMART) (Gill-Thwaites 1997), *Western Neuro Sensory Stimulation Profile* (WNSSP) (Ansell and Keenan 1989) and others. These various assessment techniques stand in the tradition of the *Glasgow Coma Scale* (GCS) (Teasdale and Jennet 1974) – a rather rough and brief method that seeks to measure neurocognitive responsiveness by means of specific stimuli and/or behavioral observation.

The SMART technique takes the search for subtle indications for consciousness even further: The juxtaposition of the terms „assessment" and „rehabilitation" in the name of this assessment technique indicates an important feature of diagnostic practice: in situations of high uncertainty the two-step logic i.e., first, to establish a diagnosis and, second, to advise a treatment, has to be revised or at least supplemented by a more circular practice. The diagnostic process starts with therapeutic actions and diagnosis follows from these actions and vice versa: to understand is to act and to act is to understand. In medical and therapeutic contexts the entanglement of diagnosis and therapy is widely accepted and a common practice. (Mol 2002)

By means of a SMART-Kit comprised of various stimuli the assessor is both observer and therapist: she or he observes potential behaviors and simultaneously intervenes by presenting a set of stimuli in order to test the patient's sensory modalities (visual, auditory, tactile, olfactory and gustatory) together with her/his abilities in motor activity, functional communication and wakefulness. The result is a structured and formalized diagnostic setting that puts its results into a hierarchy of scores (Gill-Thwaites 1997; Gill-Thwaites and Munday 1999). This setting (duration per session approx. 60–90 min) is repeated 10 times during a short time frame before any diagnosis is reached. (Our involvement in the implementation of the SMART technique at the care unit where we conduct our research is described elsewhere (Kraeftner and Kroell 2009)) Even though this procedure aims at objectifying data gathering, it accepts and even relies on a behaviorist experiment that embraces the possibility, that, eventually, it may help to generate the phenomena that it tries to measure (about the experimenters bias: see later).

Second approach: Functional imaging   The second technological development we refer to is the attempt to detect conscious awareness by looking "directly" into the head/brain. In contrast to the aforementioned SMART method that acknowledges the fact that observer and observed person may interact and thus give rise to the observed phenomenon, neuroimaging technologies attempt to avoid strictly that "averaging can produce a new quality in the average brainset that is not present in any of its source brainsets" (Dumit 2004: 86). This quote refers to the fact that the production of functional images is a complex illustrative statistical work that strives to disprove criticism that it deals with the proof of no proof. This image production (in the following we refer mainly to functional Magnetic Resonance Imaging – fMRI) is the result of a long chain of materio-technico-computational translations/transformations that starts by assuming that there exist neural correlates of consciousness (for a critical discussion see e.g. Noe and Thompson 2004, for a "theory of consciousness" see e.g. Tononi 2008). FMRI then, detects, in real time, magnetic fluctuations that are caused by an increase of oxygen signals that occur in an area of heightened neuronal activity. (The problems that are associated in correlating local neuronal activity and haemodynamics see e.g. Sirotin and Das 2009). Scanning means to measure these signal changes thereby providing a set of anatomical and functional data. Whereas fMRI provides spatial resolution of about 1–2 mm, temporal resolution is between 0.1–6 seconds which is not able to keep up with "conversations between brain areas" (Laureys et al. 2002: 159). Further links of this chain of transformations/translations are the realignment of data to compensate for movement-related confounds; the normalization of data to spatially match a standard anatomical space; and the smoothing of individual data before statistical analysis. The final chain links, then, follow two analytic approaches as there is functional segregation or functional integration: The first approach constructs "x-rays of significance of an effect, which can be projected on a three dimensional representation of the brain" (Laureys et al. 2002: 164); whereas the second approach identifies brain regions that show condition-dependent differences in the way their activity relates to activity in another chosen area. (Friston et al. 1997, Laureys et al. 2002).

In December 2005 a 23-year-old woman, who remained unresponsive for five months after severe traumatic brain injury, had to swap her dwelling and her head was put into a scanner. Of course, in the case of this woman, since no *clinical* findings showed any behavioral signs of conscious awareness, the aim was to find out if there was any neural activity that could indicate conscious awareness. Between the repetitive loud noises that in each case last for 1.6 seconds and that are part of operations of the scanning apparatus, repeatedly a single sentence or a noise equivalent was presented to the patient. First she was asked to imagine playing tennis, and then, to imagine visiting the rooms in her home. And the results were astonishing. The visualizations of the activity patterns show similarities with that of healthy individuals. (Owen et al. 2006) The images of the patient's neural activity became famous: they maintain to proof the presence of conscious awareness – by looking into the brain of an apparently misdiagnosed patient.

Experimenters Bias Revisited: Welcoming "Clever Hans"

We briefly described a few technological developments and practices that – among others – play an important role in the question of how to diagnose, treat, and care for persons in a vegetative state. But that's probably a not entirely correct phrasing of what is going on since, in fact, we witness a shift from epistemological questions to ontological ones: rather than asking how to diagnose, treat, care for a pre-given syndrome, the described (technological) developments and practices of diagnosing, treating and caring enact a different kind of syndrome or better different, multiple versions of syndromes. (Mol 2002) Of course, until recently there existed already two slightly different medical definitions, one envisaging the syndrome as a possible transition to the better, i.e. to partial recovery that justifies considerable rehabilitative efforts (Gerstenbrand 1967); or, on the other hand, the syndrome as a permanent state that questions rehabilitative efforts as futile. (Jennett 2002)

However, it is only in recent years that professionalized nursing care (Steinbach and Donis 2004), together with the aforementioned technological developments, started to increase the awareness that the affected bodies/persons might affect the carers themselves, who learn to become aware of additional versions of the syndrome. What we mean by that is, that carers interact more and more with patients on various levels and that these interactions enable both to explore the "complicated links between consciousness, affects and bodies". (Despret 2004: 114)

We took this quotation from a paper where the author investigates the case of "Clever Hans", the horse that, at the beginning of the 20th century in Berlin, was famous for its ability to solve arithmetic and other intellectual tasks. She describes the interaction of humans and the horse Hans and how "science", embodied by the experimental psychologist W. Pfungst, investigated the question of unintentional signals between Hans and humans. By referring to the isopractical behavior of a talented rider who unwittingly transforms her or his body into a horse's body, she comes to the conclusion that "both, human and horse, are cause and effect of each other's movements. Both induce and are induced, affect and are affected. Both embody each other's mind." (Despret 2004: 115)

This logic of mutual transformation of rider and horse opens, for us, an interesting perspective on the interactions between patient and caregivers: one witnesses a strange dance between human actors (Kraeftner 2008) that makes bodies and persons emerge in various ways. But why consider these interactions as a dance or articulation merely between two *living* entities? Why not include and welcome technological materiality as ingredients, as participants, as enrichments of the dance between patients and carers that thereby could result in various enactments of the syndrome, versions of consciousness, versions of bodies? Why not introduce additional technologies and material-technical artefacts and proclaim their official and important, even though not predetermined, role in the realm of traditional care practices?

But then, how to keep or increase the richness of these practices in the face of scientific research that quite recently started to investigate those people who allegedly vegetate in complex clinical-social environments? Since we know that scientific research aspires to avoid the so-called Clever Hans effect, i.e. the involuntary influence exerted by the investigator on her or his research subject during interaction: this avoidance means to isolate subjects in apparatuses or procedures with the risk of missing interesting phenomena that only arise during, in our case, clinical every-day interactions.

What is at stake? In our case, the invasion of the scientific method into clinical care may come along with the well known risk to endanger the openness towards multiple not determined articulations of (technological) artefacts and humans; with the impoverishment of clinical realities; with the debunking of "confounding" passions to clear the way for a more objective world; and with rationalist voices, critically quoted by William James, that maintain that "even if your intuition is a fact, it is not *understanding*. ... It is a mere experience, and furnishes no consistent view." (James 1912: 99)

The Scientific Method Colonizes the Syndrome?

At specialized units, up to now, nursing care and the nurse's implicit and explicit knowledge of "their" patients were the predominant sources of guiding multi-professional decision-making concerning the care for patients in vegetative state. However, there are attempts, by means of conducting nursing research projects, to investigate the effects of situated and local nursing practices, and to possibly translate those to universal and "evidence-based" knowledge that supports their qualitative and quantitative application with various patient (sub)groups. (Kraeftner and Kroell 2009) Beyond intuition and experience, assessing and measuring of possible effects, the measuring of changes of states, becomes of major importance. At the same time these measurements, when introduced into clinical practice, tend to initiate the re-ordering of articulations: nursing lines up with scientific method, medicine foregrounds the role of consciousness for diagnosis, consciousness becomes entirely mental, accessible, solitary and individual, an isolated information processing mind that is investigated in a scanner or a standardized examination. The notions of embodied, lived, materially distributed cognition or consciousness (Hutchins 1995); it's possible emergence in a complex and rich environment; the learning of becoming affected or aware as experience made by both actors, patient and carers; the multiple situated relations(hips); all of this tends to be re-assessed and displaced by the quest of *understanding* – of what's going on – in the name of the patient, of course.

Measurements of signals of conscious awareness by means of serial assessments like SMART, or functional imaging technologies, follow the rationale to avoid misdiagnosis of patients with minimal conscious state. In contrast to clinical assessments that have to take into account the Clever Hans effect, functional imaging and "coma science", by referring to their crucial experiment of detecting awareness in a misdiagnosed patient (Owen 2006), goes further and aims at banishing the Clever Hans effect entirely from the world of the interactions between patients and carers: it is about banishing the human error "when the signal-to-noise ratio (in this case, very few clear behavioral responses) is particularly small". (Monti et al. 2009: 85)

Does it mean to align with the pessimistic view that "old" or traditional technologies are displaced by "new" technologies, thereby replacing old realities with new (social) realities; to learn that caring and warm-hearted humans are replaced by cold machines or instruments; to face a repeating history that we can trace back to 19th century physiology? A scientific discipline that wondered how to replace living (human) sensors that may be able (or not) to detect or diagnose pathological variations by feeling the pulse of a person by means of the trained sensitivity of her or his fingertips; a scientific discipline that was eager to develop procedures that could exclude any error prone human mediator from measuring and thus helping e.g. the blood pressure to present "itself"? Did Ludwig's kymograph and its graphic representation of the oscillating blood pressure entirely replace the fingertips of the clinician?

A Historical Detour and a Risky Bet

However, history may tell a different story. We may learn that old technologies or instruments sometimes remain in place. Of course, it is likely that old and new technologies or instruments are incommensurable since they simply have no common measure but it is also likely that old and new technologies may live side by side. (Hacking 1992)

There is a historical example that shows an interesting coexistence of "old human subjectivity" and "new instrumental objectivity" that provides us with some important clues as to whether our methodological approach (see below) allows a realignment of objective autography and subjective assessments.

> The simplest of these machines is called 'automatograph,' and consists merely of a wooden sling suspended from the ceiling upon which rests the arm of the person being examined. A pencil is grasped lightly in the fingers, the point of the pencil just reaching a smooth sheet of paper, which records the involuntary writings. (*The New York Times* 1907: 9)

The author states that this simple machine is able to detect the involuntarily twitching of arm and finger muscles of persons being under stress "of any passing emotion." He then briefly describes the "pneumograph" which records "any variation in breathing caused by an emotional suggestion" and finally he mentions the "sphygmagraph", a device the author describes as delicate instrument that records the actions of heart beats. All these instruments are attached to a person at once and thereby producing "involuntary records" – a charted knowledge of every emotion at hand. By quoting scientists the author refers to "truth-compelling" machines that will allow distinguishing guilty from innocent people "and that a few years hence no innocent person will be kept in jail, nor, on the other hand will any guilty person cheat the demands of justice." (*The New York Times* 1907: 9) (The line of argument of the quotation resembles delicately current discussions (Monti 2009) about delineating patient-subgroups in vegetative state: by replacing the word "innocent" with "minimally conscious" and "guilty" with "vegetative" one could argue that nothing much changed during the last hundred years.)

The article displays the forerunners of the lie detector, also known as polygraph, that later would record simultaneously variations of heart rate, breathing, blood pressure, and sweating. These machines were descendants of scientific instruments developed at the renowned *Institute for Experimental Psycho*logy of Wilhelm Wundt at Leipzig, and were important elements of Wundt's former student Hugo Münsterberg's psychotechnical research agenda at Harvard. (Vöhringer 2007) Münsterberg was convinced that these machines in particular, and experimental psychology in general, would be able to support the "legal instinct and the common sense" of "the judge and the juryman." (Münsterberg 1923) These machines would help to avoid the problem that the human witness is not a source of reliable testimony. Hence, did these machines, did the polygraph – the lie detector – replace human testimony? While designed as machines to sort truth from falsehood and to avoid any Clever Hans effect, today, these machines are considered as kind of placebo-technology, where Clever Hans has entered the stage via the backdoor again as the effectiveness of the instrument depends on the intimidation skills of the interrogator (Silbermann 2006) or the subject's belief in the efficacy of the instrument. (Alder 2002)

The Jury Is the Lie Detector

Furthermore, the polygraph did not invade the province of the jury where expert testimony on issues of credibility of eyewitnesses or evidence based on polygraph tests in general is barred. (Simmons 2006) The admissibility of scientific evidence, for the U.S., was settled by the "Frye case". Sixteen years after the appearance of Münsterberg's instruments in the newspaper, in the court's opinion, the deception test had not gained enough "standing and scientific recognition among physiological and psychological authorities." (Frye v. United States 1923) Since then, and in Western countries, the polygraph is not allowed to enter the courtrooms as witness to testify the credibility of humans. There, the jury rules as lie detector. The same may hold true for functional images: The court battles over the admissibility of these images are just beginning. (Dumit 2004)

However, for a century the human and technological lie detectors have coexisted more or less peacefully: the collective human lie-detector formally accepted in court rooms, whereas the solitary polygraph, more informally, succeeded in attracting clients among police, prosecutors, corporate managers and counterintelligence authorities. (Alder 2002)

This coexistence makes us wonder if it is possible to construct a combined sensor: a technical-human hybrid that is able to detect whether patients in vegetative state are lying ... pardon ... are actually minimally conscious.

Pillow Research or "Don't just peer, interfere" (Hacking 1983)

The historical example provides us with the idea that it may be worthwhile to import a common and well-known technology that is used in jurisprudence ("the jury") into the realm of (medical) care where the scientific method and its objectifying apparatus' claim for diagnostic leadership. Following the historical example we shall believe in the co-existence of clinical and technical methods.

To summarize: the argument goes that there exist "non-behavioral" minimally conscious patients. However from another point of view, one could argue that this simply is not the case and by interrelating with those patients we learn to discriminate all kinds of different behaviors but unfortunately we do not know how these behaviors relate to the (conscious) realities of these patients. That is to say, rather than considering "non-behavior" as a pre-given precondition and starting point for the search for the metabolic activities in the brain (a search that may prevent carers from withdrawing fluid and nutrition as the argument goes), we consider this ignorance as starting point for a research project called pillow research, a certain kind of "research in the wild" (Callon and Rabeharisoa 2003) that is interested in studying how patients, carers, and we ourselves, learn to behave i.e. to be affected during everyday care practices.

Pillow research aims at the construction of pillows – in a very broad sense of the word: it is an attempt to experiment with forms and functions of sculptural medico-technical-artifacts, with "transitional objects", in the context of the clinical (nursing) routine. The notion of what a pillow is or what it can do, should be the result of a collaborative procedure at the ward where patients and non patients live together. We consider these interactions as a complex mutual process of bodily, cognitive and emotional assessments of the respective actors.

We would like to call these assessments "diagnoses", namely as an attempt to gain a (clinical) picture, impression or opinion of the state of the bodies/persons. We assume that these diagnoses are performed within „diagnostic settings" which rely on the integration of technical artifacts and therapeutic interventions, in a broad sense. Thus, we utilize the term diagnosis not exclusively for medical or nursing procedures but for all kinds of formal or informal assessments one may encounter in the clinical context that we describe.

These considerations imply, for us, as artist-researchers, to conceptualize and to build exemplary technical artifacts and apparatuses that could help to circumvent an impoverishment of clinical realities in the field of health care that we are describing. Furthermore, it gives us a mission to enter, together with our artifacts, into a friendly and hopefully fruitful competition with the described and established assessment and diagnostic technologies – and in doing so we would like to venture a bet: we bet that our clinical "research in the wild"/"pillow research" will be able to detect awareness and behaviors in "non-behavioral" patients in minimal conscious state (*and* in caregivers as well); and we bet that this research will not lag behind the results that are offered by the outlined technologies (i.e. SMART or Functional Imaging). At the end of the chapter we shall invite the reader to place his or her bet.

This confirms our ambition, in order to prevail and to win our bet, to put our efforts into the development of a bedside test for detecting awareness. Capitalizing on historical records we will develop a proprietary imaging technology.

Pillow Squirrel

In the following, first, we give an account of idea-formation that emerged from a patient's assessment, second, we describe the evolution of a so-called technical-human hybrid sensor and third, we present some preliminary experiences and results.

A Patient's Assessment

The serial neurocognitive assessment of a patient is a highly formalized procedure. It creates a standardized situation where the assessor observes potential behaviors and simultaneously intervenes by presenting a set of stimuli in order to test the patient's sensory modalities (visual, auditory, tactile, olfactory and gustatory) together with her/his abilities in motor activity, functional communication and wakefulness. It aims to find out if, and to what extent, the patient is able to perceive; whether he or she is able to show purposeful behavior or to develop a yes/no code; how many cues are necessary to wake her up and how long he is able to stay awake; etc.

Imagine you are an assessor: you are sitting quietly in front of a young man who is seated in a wheelchair. Now, you look at him for 10 seconds, subsequently, you write down any behaviors you were able to detect, then, after 10 seconds, you look again at him for 10 seconds, and so forth – until you have gathered 30 time samples and the according observations. Your are busy in observing, categorizing observations and filling in forms.

In the case of the young man, during the last ten minutes, you detected almost no behaviors. Today is the fifth session from ten and, on the form, you indicated that he showed for the most part "no movement", with the exception that he showed "chewing" one time, furthermore, he "moved his left thumb" one time, "swallowed" two times, "yawned" one time, "pursed his lips" three times, "moved his left arm" one time, "sighed" (or was it a deep inhaling?) one time, "coughed" and "sneezed" one time. You felt that it is very difficult to determine whether he sleeps or whether he looks at you via a narrow palpebral fissure. For the assessment the patient was transferred from the bed to the chair and wheeled to a quiet place where you met him to start the assessment.

What follows then is the formal assessment of the modalities by means of various stimuli. You have at your disposal: the picture of a baby, a comb, a toothbrush, cards with instructions ("close your eyes", "stick out your tongue"), other paraphernalia to present tactile, olfactory or gustatory stimuli. You start to present the various stimuli in an accurately prescribed order. The patient does not show many reactions; well, yes, when he is asked to follow the toothbrush with his eyes he seems to look up, but you are not sure. You think to yourself that he is on (SMART) level 2 indicating that he is a person in vegetative state.

And then, by walking around the patient in order to present a stimuli you somewhat awkwardly stumble across some obstacle on the floor; you quietly exclaim "ooops" by simultaneously trying to regain your balance – you look at the patient and to your surprise he is laughing. Not smiling but laughing – the same patient who did not show almost any behavior for half an hour.

This little anecdote shows the powerful Clever Hans effect: somehow, you, the neutral observer, unintentionally influenced the subject who for some reason was affected and showed an unexpected skill, talent, behavior. A behavior that the standardized, laboratory-like situation of the assessment was not able to produce and that – in case it would be reproducible on a consistent basis – would indicate that the patient actually is at least minimally conscious.

This incidence happened during a SMART assessment that was conducted by one of the members of our research group. Of course, it entailed an investigation of whether this ability of "functional communication" was reproducible. The residual assessments were enriched by slapstick interludes by which the assessor successfully entertained the patient. The results were "formally" recorded in the "additional observation" section of the forms. Further investigations showed that, in the past, the patient's sense of humor had been observed by nurses and his mother several times. These "informal" observations were added to the formal component of the assessment.

To summarize: The traditional clinical diagnostic methods until then did not show any indication that the young man is aware of himself or his surrounding. But then, by chance, the intriguing response behavior was discovered: his laughter to specific stimuli can be described as a consistent response behavior: in these moments we face a different person. (Figure 9.3)

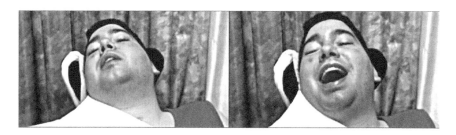

Figure 9.3   Patient watching his favorite scene

The Construction of a Technical-Human Sensor Device

This experience made us wonder whether it would be possible to produce a "diagnostic set-up" that allows to retain the "thickness" of clinical context, to accept "formal/informal", (expert/non-expert) components of information, to standardize the presentation of emotional stimuli, and, to find out what kind of stimuli the patient prefers. In the world of functional imaging one would refer to these steps as choosing a specific paradigm of stimuli presentation, data-acquisition and image processing.

Nothing much is known about the role of emotions in the world(s) of patients in a vegetative or minimally conscious state. (Bekinschtein et al. 2004; Laureys et al. 2004) Once in a while, emotional behavior (crying, smiling, laughing etc.) clinically can be observed. The relevance of these responses remains an issue of controversy. However, for us, emotions and their character as experiences that may articulate the world, the body and conscious awareness in perplexing ways, (Despret 1999) designated an interesting field to investigate. Thus it became clear to develop the design of a so called "humor pillow".

Figure 9.4   Construction of Humor Pillow v1.0

Our first prototypical humor-pillow (Figure 9.4) was a very simple device made up of a screen, a computer and a camera. The screen showed audio-visual slapstick scenes from a program comprising short scenes from e.g., The Marx Brothers, Pink Panther, Charlie Chaplin, Muppet Show, etc., and lasted approximately 30 minutes. (The selection was guided by information on the patient's biography.) The camera above the screen showed the patient's face looking at the screen and (almost) into the camera.

In order to measure any laughing activity our reasoning was rather straightforward: the signal of the camera was handled by the computer and a "motion detecting" visual tracking algorithm in a way that should trigger the very moments of laughing by capturing still images from the video of the patient's face and simultaneously from the scene that caused the laughing. The result was a double image assembled from the still image of the patient's face and the respective screenshot. (Figure 9.5) In addition, what we needed, was a kind of baseline signal, a reference signal, that would show the patient's face independently of the motion detection triggers; therefore each second we captured a still image of the patient's face.

Figure 9.5   A favorite scene

By that we were able to generate approximately 1900 – 2300 still images per 30–minutes-session. From those images we assembled one single data recording strip that showed a mixture of the regularly acquired still images and the double images that where automatically triggered and assembled by the patient's laughing. And the results were definitely encouraging. (Figure 9.6) The image showed a "humor portrait" that indicated a clear humor preference of the patient. By repeating the session several times we were able to confirm his likes and dislikes.

Confounds   This gave us the motivation to start with a second patient – a young woman – who was diagnosed as vegetative state. We compiled another "humor-

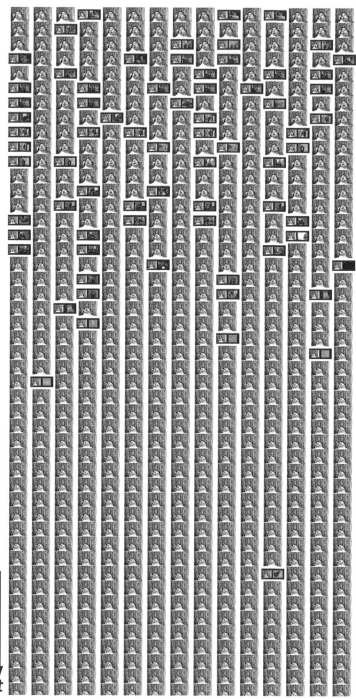

Figure 9.6   Laughing signal

program" that contained a considerable amount of funny "audio-only" scenes in order to discriminate whether the patient is almost blind (as the family member assumed) or deaf (a possibility that just as well could not be excluded); a distinction that should provide some cues whether a potential non-responsive behavior is caused by perception disorders even though the patient is in a (minimally) conscious state.

To our surprise the results were not unequivocal. We got no double images ("the signal") of a smiling patient but there were portraits that showed a somnolent, vigilant, coughing, distracted patient.

Of course, we had to learn the obvious and that "motion detection" is not equal to laughter detection: that there exists motion without laughing and laughing without motion. In addition we learned that not only the patient in relation to the camera can move – but that the camera in relation to the patient can be in motion: this happened e.g. whenever an involuntary movement hit the humor-pillow that was located in front of the patient on a little table that is fixed to the wheel chair.

We got data recordings that showed a good deal worse "signal to noise ratio" compared to the first experiment. This ratio made it impossible to establish a reliable humor portrait (Figure 9.7) To our excuse, our first patient – the young man – was almost too well suited for our "motion detection paradigm" since he almost never moved and in case he did – it was because his laughter made his head and whole body shake.

Besides the fact that *intra*session data processing showed unfeasible results, the described confounds made *inter*session comparisons impossible: i.e. it was impossible to draw any conclusions whether there were "better" or "worse" sessions in terms of frequent or rare smiles, or whether one could recognize indices that showed any effect of the repeated sessions. Furthermore it thwarted our plan according to which we wanted to re-edit our humor program so that the patients themselves, by their laughing activity, would select and generate their preferred programs themselves.

However, what cheered us up was that we found still images that made us sure that the young woman was responding, i.e. smiling to several scenes – and in fact to both: audio-only sketches and audio-visual sketches.

Before we were able to test this alternative bedside imaging technology – the humor pillow – it was necessary to discuss our endeavor with the multidisciplinary team and with family members. Previously to the experiment the simple artifact caused discussions about the structured presentation of stimuli (according to a principle of sensory regulation); the diagnostic interaction as therapeutic element in the clinical routine; and the importance and legitimacy concerning the multitude of observations – sometimes referred to as "mismatch of realities": the closer a person interacts with a patient the more behaviors he or she is able to discriminate. The question whether these behaviors are "really there" or present mere "subjective" artifacts causes constant tensions between those who are not too close to the patients (e.g. physicians, head nurses, administrators, etc) and nurses or family members who take care of "their" patients on a daily basis.

Figure 9.7   Signal-to-noise-ratio

We conducted approximately 30 humor sessions with our prototypical humor pillow. However the noisy measurements forced us to rethink data-acquisition and image processing.

First, we developed a second prototypical pillow design. It was necessary to isolate the sensitive instrument from the movements of the patient; furthermore we had to change the design so that it could be easily accommodated to the varying positions of the patient. We opted for a mobile pillow-squirrel-design with current supply that allows for meeting the various clinical constraints: patient security, reliability, etc. (Figure 9.8) The audio-visual program is presented via tablet-computer, the camera attachment was made more flexible and the video-signal is recorded in full length on a separated recording unit.

Figure 9.8   Humor Pillow v2.0: Pillow Squirrel

Second, we had to change the data acquisition module completely. Motion detection by video tracking is not appropriate for measuring smiling or laughing activity.

There were suggestions to find ways to technically discriminate facial expressions; one could think of a technical implementation of a "facial coding system" (Ekman und Friesen 1975) that attempts to conclude from the sophisticated changes of facial expressions to the inner-emotional states of a person. This approach resembles the logic of functional neuroimaging that seeks to conclude from the spatial and temporal distributions of brain metabolism to neural activity as it might be associated with conscious awareness, for example. For us, it was not about the detection of laughing as a kind of anthropological constant of a human being but to find out more about the discreet moment-by-moment changes of facial expressions of a person.

By abandoning the idea of facial coding we decided to capitalize on the aforementioned mismatch-of-realities issue and to rely on the observations made by persons who, to greater or lesser extent, were involved in the care of the respective patients. We thought of a "sensor" that combines the lie detector paradigm of the polygraph and that of the jury as the collective human lie detector: hence, by preserving the video signal sampling method we had to construct a collective assessment tool that would give various actors the opportunity to rate the images we generated during the humor sessions.

The online tool / Data acquisition   We decided to use the Internet and to create an image database that we could make accessible for registered users who should assess the images we generated during the bedside sessions. Users were asked to earmark those images of a session where they "got the impression" that the patient was smiling or laughing. By clicking on an image the impression of an individual user was saved to the database. From the 30 sessions we selected five sessions per patient, and put those sessions online. We then asked various people to take part in the assessments. For our test phase we asked family members, members of the nursing staff, physicians, and people who did not know the patients, to assess the patient images.

Image processing   We developed a simple computer program that aggregated the collected user data – in other words their impressions – and, in a next step, applied the "collective sensor-data" on each of the images by controlling the gray value of the image: 0 per cent of votes converted the image to black, 100 per cent of votes left the image without gray tint; votes in-between generated varying gray values. Finally, the images were assembled to a data strip that depicts one session of approximately 30 minutes. This was the simple image processing algorithm that we ran on our (old) image data.

Preliminary Results (Figure 9.9)

In contrast to the data strips that we received by means of video tracking we were able to improve the signal-to-noise ratio of the humor portraits substantially; furthermore, we think that there are indications of reproducible laughing signals over several sessions that designate humor preferences of patients who up to then were diagnosed as being in a vegetative state.

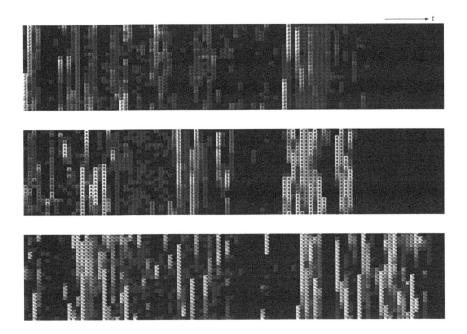

Figure 9.9 **Laughing activity patient G.M. – collective assessment: sessions 01 – sessions 03**

We think that the results will allow a quantitative comparison of sessions and to detect potential (learning or habituation) trends. What is more, the qualitative "thickness" or dimension of the patient's responses is preserved: the inscriptions make it possible to oscillate between – zooming in and out – qualitative and quantitative data components. This option facilitates any discussions on individual humor portraits thus allowing the comparison of individual observations. Of course, voting behavior of each participant can be computed and compared.

From the patient's preferences and humor profiles it is possible to build up an archive of audio-visual scenes and fragments. By establishing humor portraits of non-patients as well (multidisciplinary team, family members etc) humor preferences can be compared.

Conclusion or Better: a Risky Bet

In the near future we shall continue to test the pillow-squirrel in its bedside habitat. This is the moment, dear reader, to place your bet: do you think, that we, members of the *Research Center of Shared Incompetence* (RCSI) shall be able to compete with the "new" neuroimaging technologies in producing reliable results concerning the (emotional) awareness of patients? Do you think that barefoot technology beats high-tech science? Whom do you trust? Do you favor the machines and their inscriptions; or do you prefer the humans and their gut feeling; or maybe you can't decide and bet on the combination of both – the pillow-squirrel? Or do you tend to renounce the whole idea and think that we are simply crazy?

No, the bet is not too risky for you, dear reader – it merely asks you to take sides in the game – or to come up with new ideas to interfere. It is a little bit more risky for us because, first, our results may fail to prove that they are *no* fiction. And, second, because we feel obliged to have our hypotheses tested by those who may be directly affected by them: the actors in the field. A fact that – in case of disagreement – may lead to consequences and even to our elimination from the field.

What Is at Stake?

As artist-researchers, for us, it is not of primary concern, to support or improve, on behalf of the medical or nursing discipline, the diagnostic practice. *Pillow research* and the prototypical development of an alternative imaging technology intend to enrich the discussions about the clinical care of severely disabled people. Instead of relying on the well-known expert/non-expert, professional/lay-person, patient/non-patient, human/technical dichotomies, that tend to cut short debates on the questions of good care, our "participatory interventions" together with their (imaging) results, should create various "translations" that may lead to moments of problematization, moments that create concern and interest, that are capable of enrolling actors and keeping them in new, surprising roles. (Callon 1986)

Finally we hope to win our bet and help to keep the realm of the syndrome from being colonized by an impoverished scientific method.

References

Alder, K. 2002. A Social History of Untruth: Lie Detection and Trust in Twentieth-Century America. *Representations*, 80, 1–33.

Alder, K. 2007. *The Lie Detectors: The History of an American Obsession*. New York: Free Press.

Andrews, K. 1996. International Working Party on the Management of the Vegetative State: summary report. *Brain Inj*, 10(11), 797–806.

Andrews, K. et al. 1996. Misdiagnosis of the vegetative state: retrospective study in a rehabilitation unit. *BMJ*, 313(7048), 13–16.

Ansell, B. and Keenan, J. E. 1989. The Western Neuro Sensory Stimulation Profile: a tool for assessing slow-to-recover head-injured patients. *Arch Phys Med Rehabil*, 70(2), 104–108.

Bates, D. 2005. The vegetative state and the Royal College of Physicians guidance. *Neuropsychol Rehabil*, 15(3–4), 175–183.

Baumgartner, H. 2000. Die Ethikkommissionen und ihre Entscheidungen als Vorgabe für ärztliches Handeln, in *Medizin im Spannungsfeld von Ethik, Recht und Ökonomie: Berichte der 4. Österreichischen Medizinrechts-Tage*, edited by G. Diendorfer. Linz: Trauner Universitätsverlag.

Bekinschtein, T. et al. 2004. Emotion processing in the minimally conscious state. *Journal of neurology, neurosurgery and psychiatry*, 75(5), 788.

Bekinschtein, T. et al. 2009. Neural signature of the conscious processing of auditory regularities. *Proceedings of the National Academy of Sciences of the United States of America*, 106(5), 1672–1677.

Burns, N. and Grove, S. K. 2005. *The Practice of Nursing Research: Conduct, Critique, and Utilization* 5th ed., St. Louis Mo.: Elsevier/Saunders.

Callon, M. and Rabeharisoa, V. 2003. Research "in the wild" and the shaping of new social identities. *Technology in Society*, 25(2), 193–204.

Callon, M. 1986. Some elements of a sociology of translation: domestication of the scallops and the fishermen of St Brieuc Bay, in *Power, Action and Belief: A New Sociology of Knowledge? Sociological review monograph*, edited by J. Law. London: Routledge, 196–223.

Despret, V. 1999. *Les émotions qui nous fabriquent: ethnopsychologie de l'authenticité*. Le Plessis-Robinson France: Institut d'édition Sanofi-Synthélabo.

Despret, V. 2004. The Body We Care for: Figures of Anthropo-zoo-genesis. *Body & Society*, 10(2–3), 111–134.

Dumit, J. 2004. *Picturing Personhood: Brain Scans and Biomedical Identity*. Princeton N.J.: Princeton University Press.

Ekman, P. and Friesen, W. V. 1975. *Unmasking the Face: A Guide to Recognizing Emotions from Facial Clues*. Englewood Cliffs N.J.: Prentice-Hall.

Friston, K. et al. 1997. Psychophysiological and Modulatory Interactions in Neuroimaging. *NeuroImage*, 6(3), 218–229.

Gerstenbrand, F. 1967. *Das traumatische apallische Syndrom: Klinik, Morphologie, Pathophysiologie und Behandlung*. Wien / New York: Springer.

Giacino, J. et al. 2002. The minimally conscious state: Definition and diagnostic criteria. *Neurology*, 58, 349–353.

Giacino, J., Kalmar, K. and Whyte, J. 2004. The JFK Coma Recovery Scale-Revised: Measurement Characteristics and Diagnostic Utility. *Archives of Physical Medicine and Rehabilitation*, 85(12), 2020–2929.

Gill-Thwaites, H. 1997. The Sensory Modality Assessment Rehabilitation Technique – A tool for assessment and treatment of patients with severe brain injury in a vegetative state. *Brain Inj*, 11(10), 723–734.

Gill-Thwaites, H. and Munday, R. 1999. The Sensory Modality Assessment and Rehabilitation Technique (SMART): A Comprehensive and Integrated Assessment and Treatment Protocol for the Vegetative State and Minimally Responsive Patient. *Neuropsychological Rehabilitation*, 8(3/4), 305–320.

Hacking, I. 1983. *Representing and Intervening: Introductory Topics in the Philosophy of Natural Science*. Cambridge / New York: Cambridge University Press.

Hacking, I. 1992. The Self-Vindication of Laboratory Science, in *Science as practice and culture*, edited by A. Pickering. Chicago: University of Chicago Press, 29–64.

Hutchins, E. 1995. *Cognition in the wild*, Cambridge Mass.: MIT Press.

James, W. 1912. The Thing and Its Relations, in *Essays in Radical Empiricism*, edited by R. B. Perry. New York: Longman Green and Co., 92–122.

Jennett, B. 2002. *The Vegetative State: Medical Facts, Ethical and Legal Dilemmas* 1st ed., Cambridge University Press.

Kobylarz, E. J. and Schiff, N. D. 2005. Neurophysiological correlates of persistent vegetative and minimally conscious states. *Neuropsychol Rehabil*, 15(3–4), 323–332.

Kraeftner, B., Kroell, J. and Warner, I. 2010. The Syndrome we care for, in: *Caring in Practice*, edited by Mol, A. et al., Transcript, forthcoming.

Kraeftner, B. and Kroell, J. 2009. Washing and Assessing: Multiple Diagnoses and Hidden Talents, in *Un/knowing Bodies*, edited by J. Latimer and M. Schillmeier. Sociological Review Monographs. Oxford: Malden MA Blackwell Publisher, 161–182.

Kraeftner, B. 2009. This is (not) a syndrome. Outline of a clinico-political approach to a "consciousness-multiple", in *New Realities: Being Syncretic*, edited by R. Ascott et al. Wien/New York: Springer, 169–173.

Latour, B. 2006. Was bedeutet es, Anteil zu nehmen?, in *Die Wahr / Falsch Inc. Eine Wissenschaftsausstellung in der Stadt*, edited by M. Guggenheim et al. Wien: Facultas, 64–67.

Laureys, S., Peigneux, P. and Goldmann, S. 2002. Brain imaging, in *Biological Psychiatry*, edited by H. D'haenen, J. A. den Boer and P. Willner. New York: John Wiley & Sons Ltd, 155–166.

Laureys, S. et al. 2004. Cerebral processing in the minimally conscious state. *Neurology*, 63(5), 916–918.

Mol, A. 2002. *The Body Multiple*. Duke University Press.

Monti, M. M., Coleman, M. R. and Owen, A. M. 2009. Neuroimaging and the vegetative state: resolving the behavioral assessment dilemma? *Annals of the New York Academy of Sciences*, 1157, 81–89.

Muensterberg, H. 1923. *On the Witness Stand: Essays on Psychology and Crime*. New York: Clark Boardman Co.

Fry v. United States 1923. In *Great American Court Cases. Law Library-American Law and Legal Information*. Washington D. C. Available at: http://law.jrank. org/pages/12871/Frye-v-United-States.html [accessed July 25, 2009].

Noe, A. and Thompson, E. 2004. Are There Neural Correlates of Consciousness? *Journal of Consciousness Studies*, 11, 3–28.

Owen, A. et al., 2006. Detecting Awareness in the Vegetative State. *Science*, 313(5792), 1402.

Schillmeier, M. and Heinlein, H. 2009. From House to Nursing Home and the (Un-)Canniness of Being at Home'. *Space and Culture*, 12(2): 218–231.

Shiel, A. et al. 2000. The Wessex Head Injury Matrix (WHIM) main scale: a preliminary report on a scale to assess and monitor patient recovery after severe head injury. *Clin Rehabil*, 14(4), 408–416.

Silberman, S. 2006. Don't Even Think About Lying – How brain scans are reinventing the science of lie detection. *Wired*, 14(1), 142–150.

Simmons, R. 2006. Conquering the Province of the Jury: Expert Testimony and the Professionalization of Fact-Finding. *University of Cincinnati Law Review*, 74(3), 1013–1066.

Sirotin, Y. and Das, A. 2009. Anticipatory haemodynamic signals in sensory cortex not predicted by local neuronal activity. *Nature*, 457(7228), 475–479.

Steinbach, A. and Donis, J. 2004. *Langzeitbetreuung Wachkoma: eine Herausforderung für Betreuende und Angehörige*. Wien: Springer.

Strathern, M. 1991. Partners and Consumers: Making Relations Visible. *New Literary History*, 22, 581–601.

Teasdale, G. and Jennett, B. 1974. Assessment of coma and impaired consciousness. A practical scale. *Lancet* 2(7872), 81–84.

The New York Times 1907. Invents machines for 'cure od liars', 9.

Tononi, G. 2008. Consciousness as integrated information: a provisional manifesto. *Biol Bull*, 215(3), 216–242.

Varela, F. J. and Dupuy, J. P. 1992. Understanding Origins: Contemporary Views on the Origin of Life, Mind, and Society. *Boston Studies in the Philosophy of Science*, Vol. 130, Kluwer Academic Publishers.

Vöhringer, M. 2007. *Avantgarde und Psychotechnik: Wissenschaft, Kunst und Technik der Wahrnehmungsexperimente in der frühen Sowjetunion*. Göttingen: Wallstein Verlag.

Winnicott, D. 1997. *Playing and reality*. Reprint, London: Routledge.

Xperiment! 2005. What is a Body / a Person? Topography of the Possible, in *Making Things Public: Atmospheres of Democracy*, edited by B. Latour and P. Weibel. Cambridge Mass.; Karlsruhe: MIT Press; ZKM/Center for Art and Media Karlsruhe, 906–909.

Zieger, A. 2003. Komastimulationstherapie – was wissen wir?, *Neurologie & Rehabilitation*, 9(1), 42–45.

Chapter 10

Accessing Care: Technology and the Management of the Clinic[1]

Alexandra Hillman, Joanna Latimer and Paul White

Introduction

In this chapter we focus on our field studies of different clinical spaces in one large UK regional teaching hospital and how they perform different, yet, perhaps, complementary kinds of 'medicine': emergency, genetic and critical care medicine. We focus on those moments and processes through which patients do, or do not, gain access to the diagnostic categories, treatment and care provided by these different services. In particular, we emphasise the alignment of managerial and clinical technologies in the production and reproduction of these medical specialisms and show how gaining access to care is a critical site for exploring how and what clinical work performs in terms of mundane processes of exclusion and inclusion, and contemporary cultural and social preoccupations, such as choice (Osborne and Gaebler, 1993; Self, 1993), health (Tudor Hart, 2006) and auditability (Power, 1999; Strathern, 2000).

Managing access to health care has become an issue of intense political interest: in the UK as in other so-called neo-liberal, advanced societies, the provision of health care is both iconographic of wealth and civilization as at the same time as its funding, distribution and governance are objects of huge controversy (Brown, 2003; Alakeson, 2008). In the UK, where our regional teaching hospital is based, some aspects of health care are still free at the point of delivery, funded by general taxation, yet, as we will show, gaining access to health care is far from straight forward. Indeed, accessing health care is itself increasingly managed (Mannion et al, 2009). At the same time, medicine itself, as a system under strain, is having to find different alignments through which to maintain its place as a powerful social institution (Zola, 1972) and as at the pinnacle of the human sciences (Latimer, 2009 and forthcoming drawing on Foucault, 1976).

As we show, accessing health care is therefore better thought of in terms of a 'space' through which patients must pass in order to gain entry into the clinical domain. The nature, character and framing of this space is complex, and can be

---

1 An earlier version of this chapter was presented as 'Accessing Care: Technology and the Moral Ordering of the Clinic', *Institutions, Collaborations, Power: Workshop on Hospital Ethnography*, Sussex University, 19–20th February, 2009.

seen as a partial effect of the intensification of political and managerial agendas, including those known as clinical governance. But as we will show, these agendas, manifested and bodied forth through particular kinds of technology, need to associate (Latour, 2005) with medical agendas and identity-work for their embedding as everyday 'modes of orderings' (Law, 1994) of hospital life. It is through their associations to such agendas that these technologies become tools through which it is possible for staff in each clinical area to (re)construct and protect the production of a specific kind of medicine.

Here then we are not so much thinking about how there are different 'logics of care' (Mol, 2008) – administrative, political, medical, nursing, person-centred and so forth – that are in or out of power relations with one another. Rather we show how there are particular associations and disassociations, attachments and detachments, to and from managerial and medical technologies that afford the exclusion and inclusion of different kinds of people, work and things in the medical domains under study. It is these inclusions and exclusions that perform those domains as particular kinds of medicine. Thus, we deliberately focus in this chapter on different kinds of technology, their associations, and the kinds of relations that they call forth, in the production, reproduction and transformation of health care organization. That is we are interested in the relation between different kinds of technologies and the 'conduct of care' (Latimer, 2000).

Triage, 'Self-responsible Conduct' and the Performance of Emergency Medicine

The first study took place in the hospital Accident and Emergency (A&E) department (Hillman, 2007). The study took a particular interest in the assessment, care and treatment of older people. Participants of the study included medical staff of all levels from healthcare assistants to the clinical director and senior registrars, patients, patients' relatives/carers, managerial staff, members of Age Concern and non human actors (materials and technologies). The periods of time spent in A&E were spread across the seasons, the days of the week and times of the day and night and mirrored staff shift patterns.

A&E is understood through this work as a critical site that stands at a threshold between providing emergency care and a gatekeeper over the distribution of 'acute care medicine' as a highly valued resource. As a space in-between, between the outside world of publics attempting to gain access to emergency treatment and the inside world of emergency medical expertise, it is a space through which patients must attempt to pass (Garfinkel, 1967). Access to this domain of expertise has become increasingly protected with the fears of mounting demand for limited resources; a fear compounded by the narrowing frame within which the task of emergency medical services is understood. A&E is also a site in which the access and flow of patients must be managed, not only the attempted access of potential patients to the expertise of emergency medicine, but also the access of emergency patients to hospital wards. This task creates a space of potential conflict between the valued work of A&E (that of highly pressurised expert medical intervention with fast, measurable results), and the needs and demands of those who attempt to pass through this threshold.

A&E is continually (re)constituted as a threshold site by the daily, routine practices of staff and patients during moments of access. The triage system[2] set up to prioritise patients according to 'clinical need' is managed from within minor injuries (a section of the department located alongside the public waiting area and entrance to the department). This is a space in which staff are best able to perform a 'labour of division' (Hetherington and Munro, 1997), sorting patients out into categories of good/bad, legitimate/illegitimate, appropriate/inappropriate (Jeffrey, 1979; Dingwall and Murray, 1983).

The discourse of crisis that surround NHS provision, with continual recourse to the need to ration due to the difficulties emergency departments experience in coping with increasing demands (Steel, 1995; Brialsford et al., 2004) works to reaffirm A&E as a site of potential crisis, a space through which patients must attempt to pass in order to obtain access to emergency medicine. Furthermore, older people's attendance are often sited as one of the major factors in accounting for this crisis (Klein, 2006; Scuffham, Chaplin and Legood, 2003). This crisis discourse and the affirmation of A&E as a place for negotiations over limited resources are carried through in the social relations within the department. The distribution of materials such as the leaflets left in the waiting area[3] that provide information on triaging or the electronic notice above the waiting area that displays the waiting time are both a means of regulating as well as informing patients. These materials form patients' first experience of A&E as a threshold through which they must attempt to pass. The mediation of technologies for administering, rationing, auditing and regulating emergency medicine that permeates A&E work and shape staff patient interactions build upon and intensify this experience. Finally, the socio-spatial organisation of the service helps re-affirm that A&E is a place of negotiations (Hillman, 2007) over accessing valuable and limited medical resources.

---

2   Triage comes from the French *tri* meaning to sort and has been used to describe systems of sorting patients according to priority of need when resources are limited. Most emergency services work under some forms of triage system.

3   These leaflets serve many purposes and is entitled *Why do I have to wait? Information for patients and relatives*. The leaflets described 5 categories of patient with 1 being the most urgent and 5 being the least urgent. Category 5 is particularly interesting and reads:

Non-Urgent – Patients who are not true Accident and Emergency cases. If your injury or illness is over 48 hours old you may be advised to contact your family doctor (GP) or another more appropriate service. Although we aim to see these patients within 4 hours there may be further delay while patients from category 1–4 are seen.

A significant function of these leaflets is not only to appease patients who are waiting; they also work to regulate those patients who may fall into category five. These leaflets are not merely imparting information; there is an attempt to discipline those who are 'not true Accident and Emergency cases'.

In the following extract we present one such technology, *Jonah*. Jonah is a computer software programme that provides a checking system for every A&E patient at all stages of the assessment process. Individual members of staff who have encountered a patient at a specific stage in their assessment must log onto the system to record the assessment or treatment that has been carried out, thus automatically recording the time the activity took place. Subsequently, the length of time that passes between various stages of the assessment is also documented. For example, the time it takes for a patient to be assessed by a doctor, following a triage assessment, can be called upon by any member of staff in the department through 'Jonah'. This information can also be called upon for purposes outside of the day to day routines of those in the department, to ascertain the working practices of the unit as a whole. Through this system the progress of patients' passing through the A&E system can be monitored at all times. As Nurse Morris explains:

> It's called Jonah. It's fairly new, it's there to increase efficiency across the department. It was first introduced in Greenfields which was one of the worst Emergency Units in Britain and now it is one the best so we bought it here.

> How it works is you book patients in following their assessment and then update the system with say, SBD (seen by doctor) or R (referral) or whatever. It's the target that no patient should be waiting following assessment for more than four hours, the patient on the computer then becomes 'in breach' of the target. The system aims to ensure that everyone is made responsible for working efficiently 'cause with this, everyone is accountable 'cause it knows at all times who's responsible for each patient in the department.

> Before it was brought in it was made out like it would be this miraculous system that would change our lives. It hasn't but we are beginning to see a slight improvement from it.

The effects of the checking purposes of Jonah are significant. They create, in individual members of staff, the responsibility for their part in successfully keeping to the trust and government guidelines for patient waiting times and for processing patients through the unit as quickly and efficiently as possible. Not only must medical work be mediated by administrative materials, but the choices made and actions carried out by medical staff must be recorded, and in the process of recording them, there is an automatic recourse to the financial and temporal pressures upon each member of staff.

Interestingly the basis upon which Greenfields was deemed the best or the worst A&E department is automatically determined by the guidelines put forward by Jonah. The criteria upon which the department will be judged are both produced and tested by a technology such as the Jonah computer system. As May (2006) makes clear, 'the *promotion* of specific systems of practice are formally revealed to be 'effective', even though effectiveness is itself never politically clear cut' (p. 518). Furthermore, on the basis of Goodhart's law (1975), patient assessments are transformed by the presence and use of such auditing practices that, through a technology like Jonah, change from being measures or indicators of good practice and instead become targets. Such a change not only reduces the validity of the measure (Strathern, 1997) but also ensures that the actions and decision-making of healthcare professionals are shaped by the audit itself.

> Some governments (and the UK is an example) have discovered that if they make explicit the practices whereby people check themselves, they can ostensibly withdraw to the position of simply checking the resultant indicators of performance. Their intervention has already taken place: the social adjustment which corporations, public bodies and individual persons have already made to those self-checking practices now re-described as evidence of their accountability to the state. (Strathern, 2000: 4).

It is not a government requirement for such systems to be used by emergency departments; however the result is to regulate the behaviour of those who come into contact with them. The production of the initial assessment form, that is added to and developed by various staff members to form the patient record, is also the result of mediating technologies such as the computer system 'Jonah' used for patient tracking. These processes are all means with which patients and materials become 'inscribed' (Latour and Woolgar, 1986; Rose, 1989) and these inscriptions are subsequently read by the clinician as Berg (1996) shows in his work on the constituting of patients through the patient record. Such inscriptions are built upon, added to and changed over time; they have the effect of (re)figuring (Latimer, 2000) patients according to the organisationally, administratively and culturally definted categories of care helping staff establish their legitimacy to the service. When doctors 'collect patients'[4], they actually collect the two dimensional material inscriptions of patients produced through the patient record, not the 'three dimensional subject' (Mort et al., 2003: 273). The full person who exists as a moral demand (Bauman, 1993) is the three dimensional subject who may subsequently be attended to, depending upon the inscriptions of the patient that have been previously made and now read.

The extract below describes an object that is used as a tool during triage assessments to help better classify patients on the basis of their presenting symptoms.

> In a quiet moment I notice a red file on one of the desks called 'National Triage Presentational Flow Chart'. This file seeks to provide the signs and symptoms

---

4   This is a term used frequently by doctors on duty to refer to the process of collecting a patient's assessment form from one of the triage boxes, whether the patient is then called from the waiting area and assessed depends upon a number of other factors.

that will allow for a more accurate placement of patients into appropriate triage categories so that, as stated on the inside cover of the file, 'the more severe pathologies are appropriately triaged'. Inside the file are plastic wallets containing individual flow charts for specific presenting problems that a patient may attend A&E with. These flow charts ask a series of questions and provide possible responses. By following the responses a patient may give through this flow chart, a triage category is reached.

Significantly, the file is a material representation of a political endeavour that seeks to favour medical decisions based upon external research findings or, as part of the wider political movement of clinical governance, to provide clinical guidelines that represent an accumulation of clinical experience through processes of self auditing practices, performance indicators and other numerically decipherable data. It is this coming together of these two ways of thinking that makes medicine and management, in some cases and at particular moments, dependent upon one another (Ahmad and Harrison, 2000).

It is important to recognise that for such technologies to become embedded they must provide staff with the opportunities to perform not only their own identities as clinicians but also to perform a particular kind of medicine that represents the service as a whole. In the case of Intensive Care that is described in the next section, it is sometimes the practices of resistance to management technologies that can provide staff with the tools with which to (re)affirm the boundaries to a clinical area, thus helping to establish intensive care as a mode of organising health care work.

The computer system Jonah, as described earlier, provides a good example of the forms of self-auditing practices based around clinical guidelines, which, as part of the clinical governance programmes, attempts to steer individual clinical decision making towards a pre-comprehension of national guidelines. As with all clinical guidelines, they seek to place the management of medical provision at the centre of medical work so that they become an inherent part of the overall process of clinical decision making (Strathern, 2000; Rose, 1999). The self-regulation that occurs during these audit practices allows for the infiltration of discourses relating to efficiency, throughput, and resource management (including the resource of time) into the process of clinical decision making. These concerns are not explicitly separated and placed in contrast to the making of a clinical decision; rather through clinical governance they become embroiled in the process itself.

The difficulty in separating practices of audit and accountability from clinical assessment and decision-making has shifted the relations between healthcare professionals and patients so that policy driven and managerial concerns over time and resources are at the forefront of patient clinician interaction. This is not to suggest that political and managerial interests are dominating and reducing clinical autonomy. Rather, these technologies also assist in the production of A&E as a specific kind of medicine, one that separates out particular kinds of patients and tasks from its own definition as a distinct acute clinical specialty. These technologies can therefore become further materials with which staff undertake practices of inclusion and exclusion that (re)draw the boundaries to emergency medicine, to separate out the pollutants (Douglas, 1966) of those 'inappropriate' attendees such as the socials[5] (Latimer, 2000) or the 'acopias'[6] (Hillman, 2007). It is therefore the alignment and embedding of such technologies in the agendas of those staff who appropriate them that has shifted the ways in which patients are categorised as legitimate or not.

One significant consequence of these shifts in the relations between staff and patients, particularly in assessment processes, is that it is not only healthcare professionals who are being enrolled in the practices of managing healthcare resources; patients themselves are being made aware of their own responsibilities as 'health responsible citizens' (Hillman, 2007) to use health services appropriately. Calls are made to patients' own recourse to discourses of responsibility and duty in relation to burdened health care resources. Thus, the emergence of A&E as a threshold space means that it is not only medical staff who sort and order patients, but patients are being called to 'sort themselves out' according to their responsibilities as health care users. The problem that arises, especially for older people, is that choices regarding the avoidance of health need are not always open to everyone as not all citizens can equally perform 'preventative self care'.

Introducing Intensive Care

White's (2008) study of the Intensive Care Unit (ICU) was conducted between 2002 and 2006 during a period of amalgamation with the High Dependency Unit (late 2002) and included a period of bedding in of new organisational practices[7]. Individual patients were recruited to the study and followed across the course of their admission through to discharge from intensive care and the High Dependency Unit (HDU). Where possible, patients were also interviewed after discharge from intensive care to the general hospital wards. Intensive care staff including nurses, doctors, technicians, care assistants, receptionists, psychology staff, and managers were interviewed formally and/or informally (Spradley, 1979) about their understandings of intensive care and in relation to those patients, and crucially

---

5   'Socials' is a term used by A&E staff to describe those patients whose social circumstances are viewed in some way to negate their clinical need. For example, older people's needs are at times viewed on the basis of their age rather than disease.

6   'Acopias' is a term used by A&E staff to refer to those patients whom they deem to be in some way unable to cope emotionally, socially or psychologically who use A&E as a means of accessing social and/or emotional support.

7   Following government policy (DoH, 2000), the Intensive Care Unit became known as the Critical Care Unit after the 'merger', but for the most part the distinctions between the two areas remained and for clarity will be retained here.

family members[8], that had agreed to participate. As patients were followed through their intensive care admission, procedures and routines were documented. Such documentation provided an insight into who was permitted access to where, and at what points of time, which provided an idea of the place of staff, patients and family within intensive care.

Intensive care can be seen as characterised by a concentration of resources (both technical and personnel) and its intersection with other hospital environments, particularly as, at some point in time any other hospital department can be dependent upon intensive care services (with few exceptions). The research itself examined the reproduction of numerous agendas, such as political, fiscal and/or ethical which limit or legitimate access or 'disposal' (Foucault, 1972; Derrida, 1984; Hetherington and Munro, 1997; Munro, 2001[9]). As a consequence, what follows constitutes a tiny representation of the broader work and within this section the focus is upon those patients who gain access to intensive care irrespective of whether they are deemed appropriate 'to this place' and the policies and procedures that legitimate access and finally, how these practices help perform intensive care as a specific mode of organising health care work[10].

---

8   As it was through family members that consent was initially sought, in principle, prior to gaining fully informed consent from patients following general anaesthesia.

9   Here we draw from Derrida's notion of différance (Derrida, 1972), the interplay between difference and deference is read here as recognising difference and as a result of recognising (diagnosing) such difference, a deferral is made to dominant and dominating ways of thinking, or of ordering (Law, 1994) for example. For Foucault (1972), the notion of dispersal has been employed whereby nothing (in this case a given ordering, or idea) is ever truly disposed; they are dispersed for a period before their eventual return, not too dissimilar to the proverbial 'bad penny' of previous organisational forms (as we will demonstrate). Drawing upon the work of Douglas (1966), Munro (2001) makes visible broader cultural preoccupations of disposal where an attempt is made to deny or re*fuse*, to rubbish or *ref*use to preclude entry, or to categorise once the space is entered such as in our examination of A&E. As a consequence we refer here to ideas of difference (Hetherington and Munro, 1997) the ways in which particular agenda's render legitimate a certain ordering of difference, the deferring processes of that found wanting, such as the 'acopia', the normal rubbish (Jeffery, 1979), the social (Latimer, 2000), the crock (Becker, 1976), or simply not fitting current reconfigurations of proper order (the 'self-responsible citizen') and the eventual disposal (of ideas or persons), which of itself we argue, pre-supposes a return. The significance in our discussion of intensive care turns from persons, to the ethos of organizing, of the changing nature of managing the 'clinic' as the space of intensive care is imagined, practiced, rendered other and returns, as perhaps do those patients deemed 'not right for this place'.

10   And not just healthcare 'work', of itself, but the normative and normalising effects (Foucault, 1973; 1981) of particular modes. We wish to make explicit by modes of ordering s here, we follow Law (1994) and Bauman (1991; 2000) in that orders are never complete, but remain partial (Strathern, 2004), provisional and disruptable and are one among numerous modes of organising that can be discursively aligned with (Foucault, 1972, 1981) translated and attached to (Callon, 1986; Latour 1987).

This section of the chapter aims to highlight the ways through which policy mediates the relations of intensive care staff to patients, such as through demarcating eligibility for access. Beginning with accounts of the definition of intensive care, we trace changes of who is appropriate to this place and the ways in which staff are disciplined into figuring their workplace in a particular way. The power of policy to shape regard for those deemed inappropriate/appropriate 'to this place' (contingent upon nominal notions of propriety), and the register used to locate can be made visible through its disciplining effects (Foucault, 1991). These accounts are followed with a brief description of an inappropriate admission which will demonstrate how legitimacy is accorded to intensive care admissions as well as broader concerns over bed occupancy through those patients who unwittingly breach the intensive care order. What this section shows are the ways in which staff's appropriation, alignment or resistance to the changing policies for admission procedures helps them (re)draw the boundaries of Intensive Care as a mode of ordering itself and others. To summarise, the means through which admission procedures act as a particular technology that enable a particular regard in which patients are held when deemed an inappropriate admission will be explored.

From Intensive to Critical

Intensive care can be seen as a distinct form of healthcare practice in which location, technology and dependency characterise the space of intensive care. Rather than areas of the hospital that are configured to deal with particular pathologies or treatments (for example), intensive care is open to potentially all patients, in common with the Accident and Emergency department. However, intensive care tightly defines the criteria of who can access the space and at times colonises other hospital spaces (Carmel, 2006) even if in practice it does not always have complete control over its boundaries or on who secures access (as will be made explicit). Intensive care by its own definition exists for the sickest of the hospital population:

> '*with a potentially recoverable disease, who require more intensive observation and treatment than is available on a general ward*' (Intensive Care Society, 1997).[11]

Intensive care as a particular form of medicine is understood with regard to those granted access and as a self definition, the Intensive Care Society reinforces the emphasis of illness severity, but access is reserved for those with a 'potentially recoverable disease' unlike say the Accident & Emergency department or (as we will demonstrate) Medical Genetics, where recovery is not categorically pre-figured (or indeed expected). Specifically, intensive care can have strict control over entry

---

11   Part of this definition is also includes a patient who requires mechanical ventilation.

to its space, such as in relation to its self definition; however, such criteria of entry are well known throughout the hospital site and therefore a subversion of the system by healthcare professionals can be made in order to secure access, such as the intubation of a patient (who then demands mechanical ventilation on account of the breaching of the airway) in order to ensure access. On the other hand resuscitation, retrieval, and medical emergency teams often have intensive care staff as members (such as the hospital resuscitation team that comprises an intensive care registrar) or wholly organised and staffed by intensive care such as retrieval and the medical emergency team. Whilst the retrieval of patients generally involves the host intensive care staff going to another hospital site (particularly their intensive care units) they have little control over the retrieval site whereas the medical emergency team not only advises and treats patients across the hospital, but has an influence on the policies of wards and directorates as a result of their assessments, diagnostics and recommendations[12]. In this sense, intensive care which is based within a specialised and highly technical space comes to colonise other hospital spaces on account of its perceived expertise in dealing with the critically ill.

Table 10.1   Dependency Scoring System (Adapted from DoH, 2000)

Level 1	Patients at risk of their condition deteriorating, or those recently relocated from higher levels of care, whose needs can be met on the general ward with additional advice and support from the critical care team
Level 2	Patients requiring more detailed observation or intervention including support for a single failing organ system or post-operative recovery for those "stepping down" from higher levels of care.
Level 3	Patients requiring advanced respiratory support alone or basic respiratory support together with support of at least two organ systems. This level includes all complex patients requiring support for multiple organ failure

Whilst intensive care can be defined in relation to disease severity, prognosis and patho-physiological stability, health policy has further divided those eligible for access to intensive care services through dependency scoring (see Table 10.1). One of the consequences of this has been the virtual abolition of the term intensive care in favour of critical care (DoH, 2000) under the auspices of what is termed 'Comprehensive Critical Care'. Such a move has aimed to ensure that services meet patient need wherever patients are located (such as between High Dependency Units and Intensive Care Units), rather than specifically enter intensive care. Critically, such a move could be seen as moving goal posts in order that high

---

12   The border control and border expansions that (Carmel, 2006) refers to, have been contextualised elsewhere (White, 2008) in relation to the rise of anaesthesia *contra* surgery and the birth of the intensive care clinician.

numbers of critical care beds are seen to be available in quantitative terms, whilst maintaining the same number of physical beds through the amalgamation of pre-existing resources. Re-drawing the boundaries of intensive care in relation to the broader notion of critical care came alongside the re-categorising of care for Level 2 (in practice this meant one nurse could care for two patients) and Level 3 patients (one nurse allocated to one patient), whereas previously intensive care largely concerned itself with Level 3 patients, providing one to one care between nurse and patient. Such a terminological change required a broader organisational change; the consequences of which (we will demonstrate) altered the ways in which patients were regarded. Within the hospital ICU, staff were encouraged to use the terms critical care (and could be verbally chastised for getting the terminology wrong) and the appropriate patient dependency terminology when allocating and discussing patients among intensive care staff.

> '...so I think, you know if you accidentally let it slip you know you're gonna go to HDU rather than I'm gonna go to the south side of the unit then um you know I think, I think your forgiven the odd slip, everybody's human everybody makes the odd slip at one point or another aren't they* (From interview with Staff Nurse)

The idea of intensive care had to be removed from everyday talk to enable a new discourse of 'Critical Care' and patient dependency. Critical care patients were treated according to dependency, not in a particular space with the most appropriate resources as the resources should follow the patients (DoH, 2000) and to some extent this holds because the team has no choice but to accept this discourse, even though the idea of intensive care 'slips' out on occasion. Rather than dividing patients into appropriate or inappropriate for admission, the labour of division (Hetherington and Munro, 1997) surrounds the placing of patients into categorisations of dependency, which in turn constitutes some of the work of critical care. Rather than the vision of ensuring critical care services to be provided wherever the patient was, in practice such a policy inadvertently meant that patients were treated *in situ* but accorded a lower status (in intensive care) if a Level 2 patient and deemed an organisational risk in the High Dependency Unit if a Level 3 patient.

Whilst 'critical care without walls' was an attractive policy move in practice it was quickly discovered that the former High Dependency Unit areas were not equipped for the care of what would have been referred to as intensive care patients in terms of space or resources (staff or technology) and was quickly abandoned until a new category of patient (the long-term Level 3 patient, or 'weaner'[13])

---

13 The term 'weaner' derived from a patient that is being weaned from mechanical ventilation and often includes long term patients who have been in intensive care from weeks and months, to years in some rare cases. Whilst the term was in use throughout the period of the study, it came to be symbolised with less dependent Level 2 patients, even if the patient required mechanical ventilation.

was introduced, which again was a category of patient previously perceived as inadmissible to HDU. Such categories enabled a shorthand categorisation of patients, with Level 2 and 'weaning' patients receiving pejorative status within intensive care. For many intensive care nursing staff (although not all) in intensive care, 'caring' for Level 2 patients was seen as a waste of resources and expertise (of the nurse and organisation), to the extent that some nurses asked specifically for a Level 3 patient or not to take a Level 2 patient. On the other hand some nursing and medical staff made a particular effort to 'discharge' such patients to the 'HDU end' of the unit in order to keep them out of the space designed specifically for Level 3 intensive care patients.[14] One of the consequences of the new form of 'scoring' patients came a new way of seeing and managing patients as dependency levels created a new cultural material through which the critically ill could be regarded. This was particularly notable when nurses were allocated patients that were not deemed appropriate to this place, with many resisting allocation to such patients or making it very difficult for High Dependency staff to refuse transfer, particularly through the strategic use of intensive care consultants. Thus, through both the enrolment in and resistance to these technologies governing patient admission, staff were able to (re)affirm the boundaries to their work, their patients and their orderings of intensive care.

Organisationally, new categories of appropriate and inappropriate, admissible and inadmissible, legitimate and illegitimate were created or redrawn, with attachments, detachments, associations and dissociations made between categories of patient. To some extent the space of critical care became a space of resistance to policy, as will be shown. Although resistance was present in the practice of staff, the instrumental level of care remained the unchanged, the work of treatment and monitoring still needed to be performed, and it was not the fault of patients who by now had been recast as innocent pawns in power games between management (who enforced policy), and former HDU and ICU staff who translated the categorisations into something meaningful (Callon, 1986) within the organisation. Eventually it was decided that operationally the units should be split into three, with one nurse in overall charge of the three areas and then a nurse in charge of each (two intensive care and one high dependency) individual area, with all members of staff having to rotate on a monthly basis to each area. In effect the old system was recast, but without any managerial or organisational structure to what was the High Dependency Unit (now re-cast as a Level 2 area[15]) and whilst organisational changes may have re-drawn minor boundaries and altered notions of appropriate

---

14   This of course was not a new phenomenon as intensive care had a habit of transferring aggressive or difficult to manage patients to the HDU; however it became less easy to legitimate disposal of what were seen to be inappropriate patients once the two areas were amalgamated, as the fear of High Dependency staff was that intensive care would regard High Dependency as a 'step down unit' for long term and difficult to manage patients.

15   Spatially the High Dependency area was seen as unfit for purpose as it was not a specifically designed site for Level 3 patients.

admission within critical care, inappropriate admissions from outside critical care remained an issue.[16]

As will be seen, one of the causalities of the re-organisation was a lack of a separate management of High Dependency, as a result it became a space which was entered reluctantly by many of the medical and nursing staff and even (to a lesser extent) former HDU staff. A 'new normal' mode of ordering permeated the unit; Critical Care, complete with the *'dumping ground'* as many now fashioned HDU. It was transformed into a space where long term, aggressive or difficult to manage patients came to reside. The criteria of appropriate to intensive care became the dominant category, as we will make visible in the following example of a patient traversing 'critical care'.

On Getting into Critical Care

A patient was admitted post-operatively following the repair of an Abdominal Aortic Aneurysm (Triple A repair [17]). Generally these patients would by-pass the intensive care unit, being admitted to the High Dependency Unit from the Recovery Room of the Operating Theatre suite. Here they would be observed overnight for perfusion of their feet and of their blood pressure, they are admitted for 'bed and breakfast'[18] as it was known to both HDU and ICU staff, simply for an overnight stay before transfer back to their home ward where their bed should be kept. However, the anaesthetist reported that the patient had a massive drop in blood pressure post-operatively, required an infusion to maintain his blood pressure in addition to additional blood products as replacement for blood loss. He was sedated and ventilated and as such should come to intensive care as he was seen to be too sick to be an appropriate High Dependency admission. He arrived at half past eight in the evening, where the intensive care staff assessed him. From the perspective of the nurse who took handover from the anaesthetist, there was little reason for the patient to be admitted into intensive care as a Level 3 patient. Rather, she felt that it had more to do with having an empty recovery room and leaving no loose ends (such as a patient) at the end of the shift which would have

---

16   Although now the HDU was fully under the control of intensive care with minimal facilities for medical or surgical teams (they had lost dedicated beds) who had to book a bed in advance (if possible) and was only admissible under the direction of an intensive care consultant and/or senior nurse.

17   Generally written as 'AAA' in the medical and nursing notes. This refers to a bulging (aneurysm) of the aorta as it descends through the abdomen; ultimately the risk is perforation of the aorta which will lead to a massive blood loss and can be fatal with no surgical intervention. The observation of peripheral pulses ensures that there is no subsequent blood loss or occlusion to the artery.

18   Such bed and breakfast patients were routinely seen as part of the work of both high dependency and intensive care, but had latterly become more of a Level 2 area responsibility.

required an additional hour to prepare him for a transfer to HDU rather than him being a legitimate intensive care (Level 3) patient.[19] She felt that if he stayed an hour or so more in recovery, he would be extubated,[20] the inotropic[21] drugs would be off and he could be transferred straight to HDU as a Level 2 patient without taking up more labour and technology in an intensive care bed. The intensive care consultant who later performed the evening ward round echoed this view, the inotropic support was now off but the patient remained sedated and ventilated. The consultant suggested he remain ventilated overnight:

> *'turn the sedation off now and see what happens, if he wakes up, extubate him'.*

The patient was extubated the following morning and later discharged to the High Dependency Unit after a wash and some breakfast, a short acting sedative infusion was given as a bolus dose in order that he could be turned and washed without disturbance overnight, a technology specific to intensive care.

Questions of whether or not patients are deemed or are not deemed appropriate to this place can be made visible with regard to the perception of another's action or inaction; in this case what is considered correct for the role of an operating theatre anaesthetist. The normal rules of admission as a Level 3 patient were seen to have been flouted. For the patient, very little could be differentiated from a High Dependency admission to an intensive care admission as this was what was to be expected after surgery given that there were no alternatives points of reference. The consequences however, could be seen in relation to a period of sedation whereby the intensive care staff could better manage his treatment when he remained immobile with a short acting sedative infusion. This also had the advantage of reducing the surveillance nursing staff would be involved in as the risk of him removing his own endo-tracheal tube was minimised. He could be (and was) washed whilst under sedation and woken up (the infusion stopped) once as much of the preparatory work for discharge (documentation, washing, removal of infusions and so on) had been completed. The cost to the patient himself could be seen as marginal as he was unaware of these things, but served to get him out of intensive care as soon as practicably possible. Such discharge and patient management work however is predicated upon the notion that he was not appropriate to this place, which meant that the overriding emphasis was on disposal, of getting him out of the wrong space and into the correct space as quickly, efficiently and at a socially acceptable time (after 09:00), especially when bolus doses of short acting anaesthetic drugs ensure a smooth process. All this work, of managing this 'inappropriate' patient and their

---

19   The operating theatre shift ends at either 21:00, or with the discharge of the last patient.

20   Extubation is the removal of an endo-tracheal tube that facilitates mechanical ventilation.

21   Drugs that increase the contractility (and hence force) of the heart beat and so elevate blood pressure.

disposal, accomplishes a re-affirmation of Intensive Care as a particular ordering of care focussed on a particular kind of patient with a specific illness severity.

Intensive Care by Another Name

The intention within this section is to re-appraise how organisational issues alter the regard that health care staff have for patients as well as to show how they are enrolled in the performance of Intensive Care as a particular kind of medicine. Consequently, notions of access, disposal and the ethos of organisations (Munro and Mouritsen, 1996) and the consequences for care have emerged and re-emerged as a trope of disposability and accessibility. In this sense the aim is to highlight particular spaces of care that emerge as a consequence of particular policy moves. Both intensive care and high dependency hold a particular set of relations that delineate work, in this case for patients. However, such changes alter the relations somewhat between health care staff from different areas and between health care staff and patients. Yet amidst all this change, the previous demarcations of intensive care and high dependency work, were never truly disposed of as the old boundary re-emerged under a different name and without an independent organisational structure (in the case of the High Dependency Unit). The old distinctions remained irreducible to one another and remained un-superimposable (Foucault, 1967) as time worn practices continued and new policy initiatives were transmogrified into the cultural understandings of intensive care (Gibbons et al., 1994). At the end of the day, the intensive care staff ultimately made visible who and under what circumstances was an appropriate admission to this place.

The orderings of appropriateness to place makes visible the relations between the organisation, policy, and patients. The labour of division surrounded how patients were regarded, where the work of division involved the translation of pathology into a neat and discrete category. The effects of such (re)figuring of critical care effectively rendered marginal the space of HDU and the relative worth of those patients not deemed appropriate to intensive care. However, the power of the cultural forces that bound intensive care as a distinct mode of ordering practice ensured that the old divisions remained in place, even when new modes of ordering were implemented, intensive care as an imagining of practice and organisation could not be truly disposed. As a site of significance, the HDU was rendered marginal to the proper work of intensive care, even though the empire of intensive care expanded. However, business ultimately continued as usual, yet the regard held for patients by the organisation was altered as a result of extending into the cultural material of policy (Munro, 1996; 2005; Strathern, 2004; White, 2009). Whilst much time has been devoted to medical technologies and their impact upon care across literatures, we have tried to un-conceal managerial technologies and their effects upon persons, the accomplishments of policy in everyday mundane intensive care practice before examining the modes of organising the individual in line with genetic understandings.

Categorising Patients: Audit, Genetic Medicine and the Rebirth of the Clinic

The material presented about the practices and processes of clinical genetics comes from a study of dysmorphology. Experts in dysmorphology define themselves as concerned with the study of abnormal *forms*. Patients are mainly babies, children and their relations. The dysmorphology clinic works the boundaries between genetic techno-science, discourses of normal child development, medicine and the family. Dysmorphologists identify patterns in facial and other features, and in clinical and family histories, in the construction of what they call 'syndromes'. At the time of the study there were over 3000 syndromes recorded in dysmorphology databases. The clinical enterprise is particularly focussed on establishing whether syndromes have a genetic base.

Field work included observations of home assessments, clinical consultations and team meetings across one regional genetics service; local, national, and international academic meetings of clinical geneticists and genetic scientists at which dysmorphologists present their cases; interviews with patients or their relations referred to the genetic service; interviews with internationally renowned geneticists identified by peers as experts in dysmorphology; and examination of various 'syndrome' websites. In what follows we explore genetic medicine, as a space for the interaction of genetic science, clinical medicine and the family (Latimer et al. 2005, Latimer 2007a and b, forthcoming). So how do people access the genetic clinic? What in the genetic clinic constitutes an appropriate patient or family? And how is their appropriateness assessed? And what do these ways of assessing appropriateness accomplish?

A challenge facing the genetics clinic, as discussed in one of the academic meetings, was how to make the services that geneticists provide visible to management. What emerged from this discussion was how the auditing of clinical practice in the context of genetics was problematic: the presentation explored how the usual criteria used to measure and evaluate clinical practice did not completely fit geneticists' work. This lack of fit pertained to both the categorising, classification and diagnosis of genetic disorders and the provision of treatment and care, as well as the issue of timing and length of time that a patient and their family was on the books. Thus diagnostic categories of classification in genetics do not fit established 'disease' classification. In addition, treatment and care are difficult to measure because geneticists do not treat patients, they diagnose and counsel them with regard to reproductive risks, and diagnostic work itself is protracted and full of uncertainty.

In a sense, then, geneticists practices are solely to do with the circulation and interpretation of information, information through which they attempt to, indirectly, *reform* those families where they think there is risk of a genetic disorder being passed on. The geneticists' interest is not really with individuals, but with the extended family, across and up and down generations. As well as all the usual clinical technologies and processes, such as clinical history, biochemistry and CT scans, there are three distinctive materials used in dysmorphology diagnosis: photographs, family trees, and molecular and cytogenetic tests. Team meetings consist of all members of the clinical team working under the direction of the genetic consultant and are key occasions: the different materials produced by each aspect of the clinical process are assembled, and interpreted in the work of genetic categorizing. Here the practical work that goes into the recognition and description of dysmorphic syndromes can be observed. An extraordinary feature of this work includes explicit uncertainty, and the *deferral* of decision-making (see Latimer, 2007a). Given the context of contemporary health care and the emphasis on throughput and speed questions arise at how it is that in the context of explicit uncertainty and deferral clinicians manage to, as they put it to each other, 'keep people on'.

The key aspect then that we want to focus on are those practices around how clinicians legitimate 'keeping people on'. Examining how clinicians legitimate 'keeping people on' as patients for whom the clinic has some responsibility allows the grounds through which patients' appropriateness to genetic medicine (and their continued access to the clinic and to care) to come into view. In addition, it allows the specificities of what makes up good medicine material for genetic medicine to be described. Critically, it emerged in the analysis that in the genetic clinic the work of 'keeping people on' is as important as the work of making a diagnosis, particularly because arriving at a diagnosis was not an easy or unproblematic affair. This observation contrasts with other studies of medical processes, including those observed in A&E discussed above, that suggest that clinical work is never just a purely clinical matter, but that the organizational context mediates clinical decisions, so that clinical work is aimed at achieving decisions that lead to the *disposal* of patients (Berg, 1992; Latimer, 1997). This is not to suggest that all patients are kept on: what we see in the work of legitimating the keeping on of a case is that there are, as with A&E work, dividing practices, and mundane processes of inclusion and exclusion, at work here. Furthermore, the technologies that mediate these processes, unlike A&E and Intensive Care, are not so much those of audit and accountability. Instead, the materials that represent the families, alongside clinicians' own accounts of them, are mediated by discourses of deferral and uncertainty.

Sometimes deferral is legitimated because the patient is a baby, and 'signs' of a phenotype may emerge as the child gets older. Clinicians refer to this as 'growing into the face' of a syndrome. However, some children/families are performed as not fitting any recognized syndromes in the present but that this may change. Here a child and their relations may need to look dysmorphic, and there may need to be evidence of different chromosomal looking features and effects across different family members to help suggest that the 'problem' is in the family. But, and crucially, in the work that goes into legitimating deferral parents and other family members own commitment to finding out what is wrong is assessed: put simply the processes of inclusion and exclusion over access to the genetics clinic revolve around not just whether the 'case' represents a possible syndrome, but whether the family are committed enough to a protracted period of diagnostic work that can perform the clinic as doing knowledge work in the making.

The first point of contact between the clinic and the family after being referred to the genetics service is a home visit by a genetic specialist nurse[22]. This home visit employs a very specific kind of technology of assessment that enables the beginnings of a family 'medical' history to be assembled and a family tree or pedigree to begin to be constructed. These home assessments are then discussed at the clinic team meeting. One of the interesting aspects of these discussions was a different kind of assessment to that which Latimer was used to: not just of whether or not the child had a medical future. On the contrary the conditions being discussed are incurable, and many have poor prognoses. Rather what was being assessed was whether the child/family (the phenotype) was interesting in terms of the making of the future, a future of discovery in genetic science and technology. Here, and this is what we want to emphasise, is how the assessment was also of family member's commitment. For example, a clinical nurse in her report of her home assessment of one family stated that there was a fridge in the front garden, that the Mum was also delayed (like her child), and that they were barely coping. The clinician summed up the discussion in the phrase the family had 'too much on'. This led to ambiguity over whether to keep the family on, keep them coming to the clinic.

The gaze then of the clinic includes an assessing gaze in terms of the family's capacities for hope and desire for knowledge. This commitment emerges as a kind of capacity in the parents and the family for finding out what was going on with the child and the substance of the family. However, the commitment to pursue a genetic diagnosis in particular was not necessarily the starting point even for those parents with a desire to know. What emerges in the interview with Kevin's parent are all the oscillations that occur around the question of 'hope', the future, and the construction of 'anticipation' (see also Romain, 2008, and forthcoming):

> Interviewer: And obviously you have the experience of Dr ***(the genetics specialist in dysmorphology)?
>
> Father: And I did ask even then I said what will happen in years to come when he's older I said genetics are improving all the time, in twenty years time it'll be more advanced, because it's interesting to see programmes on telly about them doing genetics putting things in people's veins, people having trouble with their veins a genetic, something growing in your veins, put the gene in if Kevin got a problem with his chromosome 15, why can't they just give him chromosome...

---

22   Thus a remarkable feature of clinical process here is that the clinic in dysmorphology literally extends from the hospital into the home – something that most other specialisms (including geriatric medicine, child medicine, health visiting, and general practice) have been at pains to abandon. Indeed, while contemporary health care technologies are being developed to allow for people to be monitored and made accessible to the medical gaze while remaining at home, much health care technologies are aimed at reducing the need for health care practitioners to be present in the home.

Father: You do ask when you see programmes and they can advance it say in 10, 15 years time will they take a chromosome 15 and plant it in him? And would he be, I mean I'm not saying completely but will he get better?
Mother: [mutters something and laughs]

What is generated in the interactions with the technology of the genetics clinic is not so much a hope for the present child but for future generations of children. For the current child there is not much in the way of hope:

Father: Well that's Kevin, he's like that because that's what he's...he was born like that, that's what he is, but it does make you think.
Interviewer: So you always have been hoping then that...well once you were told no, could you accept that or...?
Mother: Yeah.
Father: Yeah, but there again we've had people telling us then, yeah they'll tell you that because they don't know, they're not going to tell you what they don't know.
Interviewer: Right.
Father: But in twenty years time if they do find out things, you know

The engagement with the clinic and its materials and accounts thus calls forth the possibility of a techno-scientific future, what we would like to call an immanence (Adam 2008) of hope. Now we want to stress that parents like Kevin's are at the end of a 'process of selection'. On being referred to the genetic clinic they are balanced at a new threshold, a threshold of unimagined and potential revelatory diagnostic possibility, possibilities which have huge implications for not just their family, but for the health of future generations.

What is being assessed from home visits and in materials compiled and reproduced in the team meetings is not then just the extent to which a family is genetically interesting as a possible case, although this is of course a part of it. Rather, it is parents' and sometimes even grandparents' capacity for some kind of commitment to the clinic as a site of production of knowledge about genetics. Specifically, in their interactions with the clinic parents and grandparents are exercised between a shifting ground between knowledge, uncertainty and deferral in ways that exercise their anticipation of and commitment to a future of more knowledge, more science, but also to the need to assess the risk of any reproductive act (Latimer, 2007a). It is this commitment that helps ease their access to a dysmorphology diagnosis.

Families thus have to be drawn into to the future (of their family, of other generations). This question of how and who to keep on is in a sense the opposite set of problems faced by other clinical services such as those illustrated by Hillman's study discussed above which is more about ways to keep people out (Hillman, 2008). So in this latter idea of how to keep people on arises the focus of concern for the clinicians: who (or is it what) do they 'keep' (Latimer and Munro, 2009), and who or what do they dispose of (Latimer, 1997; Munro, 2001)?

Conclusion

Our interest in this chapter has been to explore how different kinds of technology figure in the associations (Latour, 1986) and orderings (Law, 1994) of everyday health care work and examine what gets accomplished here, in terms of specific forms of orderings and processes of inclusion and exclusion. The technologies that populate the worlds we have explored range from what are easily recognised as technology, such as breathing tubes and computers, while others are better understood as 'medical technologies' such as systems for categorising patients, and methods of diagnosis, such as the medical history or examination, while others are best described as managerial technologies, such as the apparatus' designed to enable easy audit of medical practices, or those put into practice in order to support decision-making, such as triage. Indeed, some technologies are demanded in order to efface previous forms of working, yet to dispose does not rule out a re-emergence.

We have focussed particularly on how each of our sites operates at boundaries and interfaces that require processes through which people access the expertise and resources, or 'care', that the service is meant to provide, or don't. In focussing on the complex ways in which staff determine the legitimacy of patients to their service, we show how technologies focussed upon the managing of healthcare provision mediate and subsequently shift the relations between clinician and patient, whether this be through figuring patients as pawns in securing a timely discharge (ICU) or as potential claimants of limited emergency healthcare resources. We have shown how managerial technologies cannot be viewed as systems that simply impinge upon or restrict clinical autonomy. Instead, we show how these technologies, in their practical enactments, have their own affects that shift, or don't, the relations between clinical staff, and between staff and patients on the one hand, and the kinds of medicine being performed on the other. In focussing our attention on how staff accomplish the categorisation of patients as legitimate to their service or not, we show how artefacts, their interaction with individual actors and organisational routines are intermediaries that together help produce each clinical setting as a specific cultural domain. We have therefore shown how inscriptions (Latour 1986) produced through the cultural materials that make up these technologies are also available to staff members in their representations of their work, their patients and their service. We suggest therefore that it is essential to consider both staff enrolment in the wider cultural preoccupations to which these technologies are aligned and the specific situations and contexts in which these technologies are drawn upon, negotiated or resisted as well as what is accomplished as a result.

Our guess is that these processes are not unfamiliar in the context of health care organisation and reorganization world wide in that they represent processes that are a mark of health care systems, and medical professions, 'under strain' (Giddens, 1984). Critically, we think that the landscape, or as Adele Clarke (2009) would have it, the 'healthscape' that we illuminate here is a partial and provisional representation of shifting contexts of health care. In particular the demise of a Welfarist model, and the emergence of different formulations that shift responsibility for (ill)health and care onto individuals (May, 2009), calls medicine away from the bedside, not just back to the laboratory, but to a 'science of prevention' (Latimer, 2009 and forthcoming a) in the form of more and more personalised, lifestyle and molecular explanations and techno-scientific solutions for (ill)health. It is through such re-figuring's of care through technology that the emergence of a new form of regard for patients is created and whilst such figuring's satisfy a particular agenda of ensuring quality through audit practices (Power, 1999; Strathern, 2000), an accomplishment of such formulations is to create a form of governance 'at arm's length' (Foucault, 1977; 1991; Taylor, 2002) that allow a certain regard for patients to be dismissed whilst replacing such regard with an ethos of inconvenience with patients emerging within such a space as the unfortunate benefactor of a system that whilst caring for them, has to be seen to be caring for itself.

References

Ahmad, W. and Harrison, S. 2000. Medical Autonomy and the UK State 1975 to 2025. *Sociology*, 34, 129–146

Alakeson, V. 2008. Why Oregon Went Wrong. *British Medical Journal*, 337, 900–901.

Bauman, Z. 1991. *Intimations of Postmodernity*. London: Routledge.

Bauman, Z. 1993. *Postmodern Ethics*. Oxford: Blackwell.

Bauman, Z. 2000. Social Issues of Law and Order. *British Journal of Criminology*. 40(2), 205–221.

Becker, H.S., Geer, B., Hughes, E.C. and Strauss, A.L. 1976. *Boys in White: Student Culture in a Medical School*. New York: Transaction.

Berg, M. 1996. Practices of Reading and Writing: the Constitutive Role of the Patient Record in Medical Work. *Sociology of Health & Illness*, 18, 499–524.

Brailsford, S., Lattimer, J. and Tarnaras, P. 2004. Emergency and On-Demand Health Care: Modelling a Large Complex System. *Journal of the Operational Research Society*, 55, 34–52.

Brown, T. 2003. Towards an understanding of local protest: hospital closure and community resistance. *Social & Cultural Geography*, 4(4), 489–506.

Callon, M. 1986. Some Elements of a Sociology of Translation: Domestication of the Scallops and the Fishermen of St Brieuc Bay. In *Power, Action and Belief: A New Sociology of Knowledge*, edited by J. Law. London: Routledge & Kegan Paul.

Carmel, S. 2006. Boundaries Obscured and Boundaries Reinforced: Incorporation as a Strategy of Occupational Enhancement for Intensive Care. *Sociology of Health and Illness*, 28, 154–177.

Chatterjee, M.T., Moon, J.C., Murphy, R. and McCrea, D. 2005. The "OBS" Chart: An Evidence-based Approach to Re-design of the Patient Observation Chart in a District General Hospital Setting. *Postgraduate Medical Journal*, 81, 663–666.

Department of Health. 2000. *Comprehensive Critical Care: A Review of Adult Critical Care Services*. London: TSO.

Derrida, J. 1984[1972]. *'Difference' in Margins of Philosophy*, [Trans. A. Bass]. Chicago: University of Chicago Press.

Dingwall, R. and Murray, T. 1983. Categorisation in Accident Departments: 'Good' Patients, 'Bad' Patients and Children. *Sociology of Health and Illness*, 5, 127–148.

Douglas, M. 1966. *Purity and Danger: An Analysis of Concepts of Pollution and Taboo*. London: Routledge.

du Gay, P. 2000. *In Praise of Bureacracy: Weber, Organization, Ethics*. London: Sage.

Foucault, M. 1967 [1986]. Of Other Spaces. *Diacritics*, 22–27. Republished in The Visual Culture Reader, 2002, edited by Nicholas Miizoeff. London: Routledge. Available at: http://foucault.info/documents/heteroTopia/foucault. heteroTopia.en.html.

Foucault, M. 1970 [2002]. *The Order of Things: An archeology of the human sciences*. London: Routledge.

Foucault, M. 1972 [2002]. *The Archaeology of Knowledge*. London: Routledge.

Foucault, M. 1977 [1991]. *Discipline and Punish: the birth of the prison*. Harmondsworth: Penguin.

Foucault, M. 1981 [1998]. *The Will to Knowledge: The History of Sexuality Volume 1*. [Trans. R. Hurley]. Harmondsworth: Penguin.

Foucault, M. 1991. Governmentality. In *The Foucault Effect: Studies in Governmentality*, edited by G. Burchell, C. Gordon and P. Miller. Hemel Hempstead: Harvester Wheatsheaf, 87–104.

Garfinkel, H. 1967. *Studies in Ethnomethodology*. Cambridge: Polity Press

Gibbons, M., Limoges, C., Nowotny, H., Schwartzman, S., Scott, P., Trow, M. 1994. *The new production of knowledge: the dynamics of science and research in contemporary societies*. London: Sage.

Goodhart, C.A.E. 1975. Monetary Relationships: A View from Threadneedle Street. *Papers in Monetary Economics* [Reserve Bank of Australia], 1.

Greco, M. 1993. Psychosomatic Subjects and the 'Duty to be Well'. Personal Agency within Medical Rationality. *Economy and Society*, 22, 357–372.

Heidegger, M. 1993. The Question Concerning Technology. In *Martin Heidegger: Basic Writings*, edited by D.F. Krell. London: Routledge.

Hetherington, K. and Munro, R. 1997. *Ideas of Difference*. London: Routledge.

Hillman, A. 2007. *Negotiating Access: Practices of Inclusion and Exclusion in the Performance of 'real' emergency medicine*. PhD thesis, School of Social Sciences, Cardiff University.

Intensive Care Society. 1997. *Standards for Intensive Care Units*. London: Intensive Care Society.

Klein, R. 2006. *The New Politics of the NHS: From Creation to Re-Invention*. Oxford: Radcliffe [5th Edition].

Jeffery, R. 1979. Normal Rubbish: Deviant Patients in Casualty Departments. *Sociology of Health and Illness*. 1(1), 90–107.

Latimer, J. 2000. *The Conduct of Care: Understanding Nursing Practice*. Oxford: Blackwell.

Latimer J. 2009. *New Epistemologies, Libratory Ontologies? The Gene and the (post)Human*. Paper at Unfencing the Open Panel. 100th Anniversary Conference The Sociological Review. Imagining the Political/ The Politics of Imagination. Warwickshire. 2–3 June 2009.

Latimer J. [forthcoming] Rewriting Bodies, Reportraiting Persons? The Gene, the Clinic and the (post)Human. *Social Studies of Science* (under review).

Latour, B. 1987. *Science in Action: How to follow scientists and engineers through society*. Massachusetts: Harvard University Press.

Latour, B. 1986. The Powers of Association. In *Power, Action and Belief. A new sociology of knowledge? Sociological Review monograph* 32, edited by J. Law. London: Routledge & Kegan Paul, 264–280.

Latour, B. 2005. *Reassembling the Social: An Introduction to Actor-network-Theory*. Oxford: Oxford University Press.

Latour, B. and Woolgar, S. 1986. *Laboratory Life: The Construction of Scientific Facts*. Princeton N.J: Princeton University Press.

Law, J. 1994. *Organising Modernity*. Oxford: Blackwell.

Mannion, R., Harrison, S., Jacobs, R., Konteh, F., Walshe, K. and Davies, H.T.O. 2009. From Cultural Cohesion to Rules and Competition: The Trajectory of Senior Management Culture in English NHS Hospitals, 2001–2008. *J R Soc Med.* 102, 332–336.

May, C. 2006. Mobilising Modern Facts: Health Technology Assessment and the Politics of Evidence. *Sociology of Health and Illness*. 28, 513–532.

Mol, A. 2008. *The Logic of Care: Health & the Problem of Patient Choice*. London: Routledge.

Mort, M., May, C. and Williams, T. 2003. Remote Doctors and Absent Patients: Acting at a Distance in Telemedicine. *Science, Technology and Human Values*, 28(2), 274–295.

Munro, R. 1996. The Consumption View of Self: Extension, Exchange and Identity, in *Consumption Matters: The Production and Experience of Consumption. Sociological Review Monograph*, edited by S. Edgell, K. Hetherington and A. Warde. Oxford: Blackwell.

Munro, R. 1999. The Cultural Performance of Control. *Organization Studies*. 20, 619–640.

Munro, R. 2001. Disposal of the Body: Upending Postmodernism. *Ephemera*, 1, 108–130.

Munro, R. 2005. Partial Organization: Marilyn Strathern and the Elicitation of Relations. The Sociological Review, 531, 245–266.

Munro, R., Mouritsen J. (eds). 1996. *Accountability: Power, Ethos and the Technologies of Managing.* London: Thomson International Business Press.

Munro R. 2001. Disposal of the Body: Upending Postmodernism. *Ephemera*, 1(2), 108–130.

Osborne, D. and Gaebler, T. 1993. *Reinventing Government: How the Entrepreneurial Spirit is Transforming the Public Sector.* New York: Penguin.

Power, M. 1999. *The Audit Society: Rituals of Verification.* Oxford: Oxford University Press.

Rose, N. 1999. *Powers of Freedom: Reframing Political Thought.* Cambridge: Cambridge University Press.

Rose, N. 1989. Individualising Psychology, in *Texts of Identity: Inquiries in Social Construction*, edited by J. Shotter and J.K. Gergen. London: Sage, 119–213.

Scuffham, P. Chaplin, S. and Legood, R. 2003. Incidence and Costs of Unintended Falls in Older People in the United Kingdom. *Journal of Epidemiology and Community Health*, 53, 740–744.

Smith, G. and Nielson, M. 1999. ABC of Intensive Care: Criteria for Admission. *British Medical Journal*, 318, 1544–1547.

Self, P. 1993. *Government by the Market? The Politics of Public Choice.* Basingstoke: Macmillan.

Steel, J. 1995. Inappropriate – the Patient or the Service? *Accident and Emergency Nursing*, 3, 146–149.

Strathern, M. 1997. 'Improving Ratings': Audit in the British University System. *European Review*, 5, 305–321.

Strathern, M. 2000. New Accountabilities: Anthropological Studies in Audit, Ethics and the Academy, in *Audit Cultures: Anthropological Studies in Accountability, Ethics and the Academy*, edited by M. Strathern. London: Routledge, 1–19.

Strathern, M. 2004. *Partial Connections – Updated edition.* California: Alta Mira.

Taylor, A. 2002. 'Arm's Length but Hands On'. Mapping the New Governance: The Department of National Heritage and Cultural Politics in Britain. *Public Administration*, 75, 441–466.

Tudor Hart, J. 2006. *The political Economy of Health Care: a Clinical Perspective.* Bristol: Policy Press.

White, P. 2008. *On Producing and Reproducing Intensive Care: The Place of the Patient, the Place of the Other.* Unpublished thesis. School of Social Sciences: Cardiff University.

White, P. 2009. Knowing Body, Knowing Other: cultural materials and intensive care, in *Un/knowing Bodies*, edited by J. Latimer and M. Schillmeier. Oxford: Wiley/Blackwell.

Zola, I.K. 1972. Medicine as an Institution of Social Control: the Medicalizing of Society. *The Sociological Review.* 20(4), 487–504.

Index